Not Exactly

Not Exactly
IN PRAISE OF VAGUENESS

KEES VAN DEEMTER

OXFORD

UNIVERSITY PRESS

Great Clarendon Street, Oxford O X 2 6D P

Oxford University Press is a department of the University of Oxford.
It furthers the University's objective of excellence in research, scholarship,
and education by publishing worldwide in

Oxford New York

Auckland Cape Town Dar es Salaam Hong Kong Karachi
Kuala Lumpur Madrid Melbourne Mexico City Nairobi
New Delhi Shanghai Taipei Toronto

With offices in

Argentina Austria Brazil Chile Czech Republic France Greece
Guatemala Hungary Italy Japan Poland Portugal Singapore
South Korea Switzerland Thailand Turkey Ukraine Vietnam

Published in the United States
by Oxford University Press Inc., New York

First published 2010

British Library Cataloguing in Publication Data

Library of Congress Cataloging in Publication Data

Library of Congress Control Number: 2009941589

Typeset by SPI Publisher Services, Pondicherry, India
Printed in Great Britain
on acid-free paper by
Clays Ltd, St Ives plc.

ISBN 978–0–19–954590–2 (Hbk.)

1 3 5 7 9 10 8 6 4 2

PREFACE

Vagueness is the topic of many scholarly books for professional academics. This book hopes to reach a much broader audience, by focusing on the essence of an idea rather than its technical incarnation in formulas or computer programs. Most chapters, with the exception of Chapters 8 and 9 in the middle part of the book, are self-explanatory and should be readable in isolation. In a few areas of substantial controversy, I have used fictional dialogues—a tried and tested method since Plato's days—to chart the issues. Although some things have been too complex to yield willingly to this informal treatment, it has been a delight to discover how much complex material can be reduced to simple ideas. On a good day, it even seems to me that some themes are best explored in this informal way, free from the constraints of an academic straitjacket.

The informal style of the book does not necessarily make it an easy read. For one thing, it requires a somewhat philosophical spirit, in which one asks *why* certain well-known facts hold. Moreover, we will not be content when we, sort of, dimly understand why something happens: we will ask how this understanding can be given a place in known theories. Essentially this means that we will try to understand vagueness in a way that is compatible with everything else we know about the world. It should also mean that the book itself is *not* vague, except where vagueness comes to the aid of understanding.

Existing books on vagueness, written for philosophers, computer scientists, or linguists for example, often focus either on the sorites paradox or on fuzzy logic. The present book deals with both of these topics, and with much else besides. Part I discusses a variety of areas where vagueness is difficult to avoid, by putting the spotlight on some corners—in all walks of life—where it lurks unexpectedly. Part II presents theories that aim to shed light on the meaning of vague expressions, often as a response to the sorites paradox. This middle part of the book is for the theoretically inclined: others may want to give it a miss and move on to the next part. Applications in artificial intelligence play a role throughout the book. Part III puts these applications centre-stage and uses them to ask why and when it is a good idea to use vagueness 'strategically' in communication.

Although the subject matter of this book has seldom formed the core of my work, it has engaged me intermittently over several decades, during which time I have worked in different research environments in Scotland, England, the Netherlands, and the United States. Its long incubation period has caused the book to be influenced by a large and varied set of people, some of whom are a considerable number of handshakes removed from each other, as long as my own are excluded. I had not foreseen how enjoyable and enriching it would be to exchange opinions with people from different backgrounds on a book of this kind. Let me try to do justice to a number of people's roles without writing another book.

A few friends and colleagues have stood by me during the writing of the book and helped to shape it by their comments and suggestions at various stages. I am thinking particularly of Graeme Ritchie, Juta Kawalerowicz, and Hans van Ditmarsch, each of whom must have suffered considerably under its early deficiencies, although they were too kind to show it. Invaluable comments were also obtained from Johan van Benthem (my former Ph.D. supervisor), Albert Gatt (once my Ph.D. student, now very much more), and from my colleagues Ehud Reiter, Chris Mellish, and Richard Power, each of whom has exerted substantial influence on my thinking. I am much indebted to Paul Piwek and Emiel Krahmer (perhaps my most fruitful collaborators so far), to Louis ten Bosch and Hugo van Leeuwen (for reminding me that Eindhoven's Institute for Perception Research was a special place), to Peter Baumann (whom we in Aberdeen sorely miss), Har Golsteyn (for advice on matters of measurement), Joseph Halpern (for feedback on Chapter 8), and finally to George Coghill, Norbert Driedger, Paul Égré, Raquel Fernández, Roger van Gompel, Frank Guerin, Imtiaz Hussain Khan, Richard Kittredge, Marianne Korpershoek, Wufaldinho Kudde, Margaret Mitchell, Arie Molendijk, Gökçen Özal, Dave Ritchie, Jurgis Skilters, and Derek Sleeman; to Jacobus, Roelof, and Marieke van Deemter; and to Robert van Rooij, Ielka van der Sluis, Frank Veltman, Yaji Sripada, Ross Turner, Sandra Williams, and Hetty Zock, for many other stimuli that were greatly appreciated. I thank Latha Menon and her team at Oxford University Press, for guiding me almost painlessly through the publication process. More than anyone else, I thank my dear Judith for tolerating my devotion to an activity as frivolous, as well as anti-social, as book writing.

Authors who thank others often assure us (for reasons that escape me!) that all remaining errors are their own. But surely, so a famous argument goes, these authors would never commit anything to paper that they did not believe to be true; why are they apologizing for errors that they don't believe they have committed? Even though I have no doubt that the present book contains its share of errors, I am confident that the approach to vagueness defended here will allow its readers to see through the fallacy that underlies this argument, which is sometimes known as the Paradox of the Preface.

http://www.csd.abdn.ac.uk/~kvdeemte/NotExactly

KvD

Aberdeen
July 2009

'Private investigator, huh?', he said thoughtfully. 'What kind of work do you do mostly?' 'Anything that's reasonably honest,' I said. He nodded. 'Reasonably is a word you could stretch. So is honest.' I gave him a shady leer. 'You're so right,' I agreed. 'Let's get together some quiet afternoon and stretch them.'

RAYMOND CHANDLER, *The Little Sister*

A barometric low hung over the Atlantic. It moved eastward towards a high-pressure area over Russia, without as yet showing any inclination to bypass it in a northerly direction. The air temperature was appropriate relative to the annual mean temperature (...) In a word that characterizes the facts fairly accurately, even if it is a bit old-fashioned: It was a fine day in August 1913.

ROBERT MUSIL, *The Man without Qualities*

CONTENTS

CONTENTS

CONTENTS

LIST OF FIGURES

LIST OF FIGURES

Prologue

The world may be best measured in terms of neatly quantifiable entities such as millimetres, grams, and millibars, but we often speak more loosely.

The weather, for example, can be assessed by measuring the temperature in Fahrenheit or Celsius, the atmospheric pressure in millibars, and so on. Yet, this morning's weather report is likely to speak of a cold day and, if we're unlucky, another low-pressure zone. Categories such as 'cold' and 'low-pressure' are not sharply delineated but *vague* around the edges: some days are cold, others warm, but somewhere in between it can be unclear whether a day counts as cold or not. This book asks why vague concepts—concepts that allow *borderline* cases—play such an important role in our lives, and discusses various explanations for this fact. We shall see that vagueness is inherent in all our dealings with the world around us. Vagueness may be likened to original sin: a stain that can be diminished but never removed (Fig. 1). We shall also argue, however, that vagueness is sometimes a virtue. There are often excellent reasons for avoiding precision. This view has practical consequences. It follows, for example, that the 'intelligent agents' that are being built in artificial intelligence laboratories will gain in usefulness once they manage to use vague concepts judiciously.

The mathematician Georg Kreisel is famous for having argued that informal argumentation, instead of meticulous proof, can sometimes be a mathematician's most powerful tool. The main thesis of the present book might be seen as a remote echo of Kreisel's call for informal rigour:[1] sometimes, one just has to be sloppy. In defending this claim, we shall not only discuss colloquial conversation, where sloppiness is only to be

FIG. I *Original Sin,* by Tintoretto

expected; we shall also be concerned with the exchange of serious factual and scientific information. If we can come to understand why vagueness pervades even such fault-critical situations then we shall have achieved a lot. In the same spirit, the book will often focus on relatively simple things. Where complex notions are involved—like justice, beauty, or happiness, to name but a few—the very idea of precision is hard to imagine.

This book is an attempt to account for the role of vagueness in our lives. This means that we shall ask such questions as: 'Why do people make such frequent use of words whose meaning is difficult to pin down?' and 'What do these words mean?' 'Why is it that their meaning varies so much from one context to the next?' 'Are all vague concepts basically alike in all these respects, or are there important differences between them?' Finally, we shall ask 'If we were to build a robot that can communicate, how precise would we like it to be when it speaks to us?' These questions will touch on many academic disciplines, from symbolic logic and game theory to computing science and biology, and from linguistics and legal theory to medicine and engineering.

This book is full of examples. Describing people comes naturally to us, which is why quite a few of these examples will be about people. An American friend (who, like me, is exceptionally tall) once pointed out that, in his social circle, height is just about the only aspect of the human body about which one can talk freely. To identify a person as 'an old woman', 'the bald guy', or 'the skinny girl over there' would be frowned upon; to refer to someone by his skin colour would be almost unthinkable. Height alone—according to my cautious friend—is safely neutral. I do not know whether his claim is correct in all particulars, but his point is well taken. Where I can, I shall avoid offence. Where I fail, I hope to be forgiven by readers of all descriptions.

1

Introduction
FALSE CLARITY

A few years ago, the BBC carried a news story entitled 'Students feel unsafe after dark'. To a sceptic, this header might conjure up images of nocturnal awakings from exam-fuelled dreams. As it turns out, the story, which was essentially identical to reports in other news media, was of a more serious nature. *The Times Higher Education Supplement*, which was close to the source of the story, wrote 'The research (...) reveals that while students generally feel secure during the day, fewer than four in ten feel safe all of the time (...)' (16 April 2004). Worries about safety can be justified, yet one wonders what to make of a report of this kind. The research itself may be valid[1] but by writing as they did, the journalists made it sound as if feelings of safety are an all-or-nothing affair. Yet feelings come in degrees, and it is doubtful that there is a generally agreed point where a feeling of safety suddenly turns into one of unsafety. For this reason, percentages are arguably meaningless.

Feelings may be particularly difficult to quantify, but that's not the point. The point is that numerical information is often taken for granted when inquisitive questions should have been asked. Authoritative sources inform us that the year 2004 saw an average of 200 mass disturbances every day in China, but no one tells us what makes a

4

mass disturbance.[2] We lament high incidences of failing schools, violent crime, and obesity, without questioning the norms that underlie these assessments. Yet, before you can count something, you have to know what you are counting. Consider obesity, for example: we are told that 40 per cent of British children are obese these days, and we swallow the information. Yet no matter how slim children are, one could *always* put the threshold for obesity at such a point that the percentage of obese children is 40 per cent. If someone told you that 40 per cent of people are tall, you would not take him seriously. Obesity may be different, but why exactly?

It has been observed that people have a tendency to paint reality in black and white, rather than richly varied grey tones. We seem to like clarity so much that we see it where it does not exist. The terms 'black' and 'white' themselves are an interesting example, particularly when applied to skin colour. We are all familar with predictions saying that this or that city in Western Europe will, in this and this future year, have a majority of black inhabitants. Such figures are seldom called into question, in my experience, by enquiring just how dark one's skin has to be to count as black. The biologist Richard Dawkins has a catchy phrase for the sleight of hand that allows us to think in black and white—both literally and figuratively—calling it the tyranny of the discontinuous mind. We shall soon see what drove him to this characterization.

The phenomenon is not limited to the social arena: the very corner-stones of our thinking are affected. The important notion of *causation* is a case in point. Court cases have been fought over the question whether smoking causes cancer, for example, even though the likelihood of cancer is affected by many different factors, so the notion of a cause is problematic. Smoking is a very strong factor, and this could arguably justify the simplification. Other cases, however, suggest a genuine lack of awareness that causation comes in degrees. Michael Blastland and Andrew Dilnot, in their recent book *The Tiger that Isn't* discuss cases where scientists are reported to have found 'the gene for' multiple sclerosis or asthma. In fact, however, the gene is present in a percentage of the people suffering from the disease that is only slightly higher than in the general population. Simplifications of this kind can be misleading, because they overstate the importance of a discovery.

Other authors in the (broadly speaking) popular science domain have made similar observations. John Allen Paulos, for instance, targeted the notion of food safety. He reports on a clause of the American FDA Act of 1958, which requires that 'no food additive shall be deemed safe if it is found … to induce cancer in man or animal', but without specifying a minimum allowed level of each of the relevant substances or a clarification of the word 'induce'. We like to pretend that there is a sharp division between substances that are safe and ones that are poisonous[3] but, in reality, there is only a continuum, with water at one extreme perhaps (because it takes many litres, swallowed in quick succession, to kill a person), chemical weapons at the other, and things such as salt and alcohol somewhere in between. Once again, we think in black and white, whereas reality is subtly shaded.

Examples of what Blastland and Dilnot call *false clarity* are easy to find in history and geography as well. It is often thought, for example, that expressions such as 'the Great Wall of China' and 'the Silk Road' denote well-defined entities. Yet the former only denotes a loosely delineated group of spatially separated walls which are extremely different from each other in terms of their structure, height, and age. One person's Great Wall is not necessarily someone else's, because it is by no means obvious which walls are part of the great one. Similarly, 'the Silk Road' denotes a diffuse network of paths all of which were important before ships replaced camels as the vehicles of choice for travel between East and West. In both cases it would be easy to point at boundary cases: a path, for example, that was used only during severe winters. The idea that there is a definite thing to which phrases such as 'the Silk Road' refer is illusory.

It is sometimes useful to simplify. In some areas, extremes are the norm: by and large, water is either frozen or fluid, and a person is either dead or alive (but see Chapter 4), male or female. In such cases the world around us supports our taste for clarity. But we bring the same attitudes to situations where nuance is vital. In the middle part of the book, I shall argue that people's tendency to think in terms of all or nothing may be responsible for the unproductive attitude that some students of language and communication have brought to vagueness. For now, let me note another, more dangerous consequence of this tendency, relating to public life. The point is that false clarity leaves us open to manipulation, as when politicians or others redefine words for political gain. In the 1980s, for example, Margaret Thatcher's successive Tory governments are reputed to have redefined the socially crucial notion of 'unemployment' dozens of times, narrowing it further and further, thereby allowing the figures to be polished as more and more people lost their jobs. If we were more aware that concepts such as unemployment are not cast in stone then leaders might be less tempted to lead us up the garden path.

Politicians are not on their own in the deception game: sales people must surely count as the champions of the genre, and even academics join the game when reporting about their research and begging for funding. Research these days, in case you haven't noticed, is invariably 'excellent', 'innovative', and 'world-leading'. Similarly, we are all used to hearing consumer products being recommended for being 'powerful' (when it's a vacuum cleaner or a car engine), 'healthy' (when it's food), 'fast' (when it's a car or a phone), or 'excellent value' (when it's pretty much anything), even though it would be very difficult to test such claims, because the words are essentially undefined. It is easy to see why such claims are left vague: if you claim that your product is better than some particular alternative, in some well-specified respect, you might be proved wrong, with potentially unpleasant consequences. What is remarkable is that meaningless claims are nevertheless thought to be persuasive. Why else do companies pay good money for them?

BBC News, Wednesday 26 November 2008

APPLE MADE TO DROP IPHONE ADVERT

An Apple iPhone advert has been banned by the advertising standards watchdog for exaggerating the phone's speed. The advert boasted the new 3G model was 'really fast' and showed it loading internet pages in under a second. The Advertising Standards Authority (ASA) upheld complaints by 17 people who said the TV advert had misled them as to its speed. (…) After upholding the viewers' complaints, the ASA said the advert must not appear again in the same form. Apple said its claims were 'relative rather than absolute in nature'—implying the 3G iPhone was 'really fast' in comparison to the previous generation—and therefore the advert was not misleading. The company also said the average consumer would realise the phone's performance would vary—a point they said was made clear by the text stating 'network performance will vary by location'.

The following chapter will flesh out one case of false clarity: the clarity associated with the biological notion of a species. But before we go there, let us define our theme more precisely and look ahead towards the rest of the book.

Vagueness

This book uses the word 'vague' in a specific sense: a sense of the word that is common in academic writing but less so in ordinary conversation. Let me explain.

A concept or word will be called vague if it allows borderline cases. 'Grey' is a vague concept, for instance: some birds are grey and some are not, but others are in a borderline area between grey and not grey. Good people may differ on whether to call them grey or not. The fact that such grey-*ish* birds can exist makes 'grey' a vague concept. Words such as 'large', 'many', and 'few' are all vague for the same reason. 'Fewer than five', by contrast, is *not* vague, because it creates a sharp division between two classes of numbers: the ones below five and all the others. To say that I have fewer than five children may perhaps be a little evasive (because I don't tell you have many children I do have) but this is not what we call vague. Until further notice, we shall use the word 'vagueness' to denote expressions such as 'grey', because they allow borderline cases. (Expressions that are not vague will be called *crisp*, for lack of a better word.) It is only in Chapter 3 that this standard use of the word 'vagueness' will be challenged. There is nothing judgemental in this use of the word 'vague': by calling a word vague, we are not judging whether it is a good or a bad word, a useful or a useless one.

In the course of this book, it will become clear that vagueness is everywhere: if you believe a concept to be completely crisp, then examine it more closely and it will often prove to be vague. Size-denoting terms such as 'small' and 'large' are obviously vague, for example, but so are colour terms, at least in ordinary language, where sharp boundaries are not artificially imposed on them. This is no accident of English. Even when artificial languages such as Esperanto are created and taught, little is done to explain the meaning of vague expressions, other than by translating them to words in existing languages. The relevant sense of *granda*, in Esperanto, translates to English 'tall', for example, but whether this makes a man of 175 cm *granda* is left to the imagination of the learner.

If we take a closer look at vague words then something awkward can be observed right away. Consider 'tall', for instance. In accordance with what we just saw, people can be divided into three categories: clearly tall, clearly not tall, and borderline tall. So far, so good. But as philosophers such as Crispin Wright have pointed out, this is only the start of our troubles, for it is not as if the thresholds between these three categories are cast in stone.

To see the problem, suppose you asked a number of people to point at a place on the wall of your living room, whose height separates the 'clearly tall' men from the 'borderline tall' ones. Surely, people would point at different heights. Our original problem therefore repeats itself at a higher level: just as it can be doubtful whether a man is tall or not tall, it can also be doubtful whether a man is tall or borderline tall, and so on. So although one might *imagine* a language where vagueness involves the existence of borderline cases and nothing worse, in the languages that we actually speak, borderline cases indicate the likely existence of harder problems, associated with the term *higher-order vagueness*. You can think of borderline cases as a symptom of a disease: if you are coughing up blood then that's an unpleasant experience in itself, but additionally it could be a sign of something worse. The 'something worse' indicated by vagueness is higher-order vagueness. Even though we shall rarely use the term, it will become clear that many of the problems associated with vagueness are actually caused by higher-order vagueness.

The aim of this book is to explore how vagueness works, and why it pervades communication. It is part and parcel of this enterprise to ask why vagueness is not always a bad thing: we shall see that sometimes vagueness is simply unavoidable, while on other occasions vagueness is actually preferable to precision. We shall also devote considerable space to discussing the implications of our findings for the construction of Artificially Intelligent systems, which are slowly but surely starting to be endowed with a human-like capacity to produce and understand ordinary language.

But vagueness does not only pose practical problems: it also poses difficult theoretical challenges.

Paradox

In October 2007, the List Universe, a website devoted to top ten lists, voted the sorites paradox, also known as the Paradox of the Heap, to be one of the ten greatest unsolved problems of science. Other problems included the existence of black holes, the cause of the Great Depression, and the chemical origin of life. The mechanism through which the list was composed is not known to me, but I do not contest the weight of the problems posed by the sorites paradox and its relatives, which will play a central role in this book. I will introduces sorites informally here; a more detailed discussion follows in later chapters.

The sorites paradox

In the sixth century BC Eubulides of Miletus described the following puzzle, centring around the vague word 'stoneheap' (Greek: *soros*). One stone does not make a stoneheap, Eubulides observed. But if something is too small to be a stoneheap, you cannot turn it into a stoneheap by adding just one stone. Clearly then, two stones do not make a stoneheap either. But by the same reasoning, nor do three stones, and so on. Consequently, no finite number of stones can ever make a stoneheap. The gist of the puzzle has nothing particularly to do with stoneheaps, of course. The same argument will 'prove' that there cannot exist any tall people or that you can eat as many painkillers as you like, for example.

Because this paradox lies at the heart of some of the things that are most puzzling about vagueness, we shall look at some of its forms in detail. And because it is a challenge to a mathematical approach to reasoning, known as *classical logic*, our story will require a modest amount of mathematical symbolism.

I was educated in a hard-nosed research tradition in which language is analysed as a means for expressing statements about the world. Human beings, in this view, are an awkward species who have unfairly man-oeuvred themselves in between language and the world. But in the case of vague language, it is hard to say much about language without talking about our *perception* of the world as well. To see what I mean, consider a variant of Eubulides' paradox, located in the laboratory of a manufac-turer of hi-fi music equipment. The human ear has limited sensitivity: some sounds are too soft to be audible. Likewise, the ear has limited *resolution*: a difference of 0.5 dB (decibel) is almost certainly too small to be perceived by anyone.[4] Now a modern version of Eubulides' paradox is obtained thus: a sound with a loudness of −30 dB is too weak to be audible; if it is amplified to -29.5 dB, it must still be inaudible, since 0.5 dB cannot make the difference; but then a further amplification to -29 dB must also be inaudible, and so on, ultimately implying that even a sound of 150 dB—well above the average person's pain threshold—must also be inaudible. Eubulides has struck again! This version of the paradox is harnessed by science: we actually know that a difference of 0.5 dB is undetectable.

A curious difference between scientific disciplines is worth mentioning here: by and large, acousticians are not too perturbed by sorites, because they do not care too much about the meaning of a word such as 'audible': they are happy if they can model a person's hearing with reasonable accuracy, and their models do not require them to define this concept precisely. For students of language, cognition, and communication, however, the paradox is harder to push aside, since they earn their upkeep by building models of the meaning of words, sentences, and so on, and these models thrive on binary distinctions such as the one between audible and inaudible. This difference in attitude between academic disciplines will be further explored in Chapter 9.

The sorites paradox plays tricks on us every day. The philosopher Dorothy Edgington (1992), for example, hinted at the *mañana* paradox (if a task can be postponed by n days then it can be postponed by $n + 1$ days) and the dieter's paradox (one more sweet won't do me any harm). The computer scientist Merrie Bergmann cites the example of population growth: a population growth of 0.01 per cent does not endanger living standards, therefore population growth is nothing to worry about. Similar arguments can be made about environmental pollution and climate change. All such arguments have the same structure. A good collection of examples can be found in a recent bestseller by Penn and Zalesne, entitled *Microtrends*, even though sorites is not mentioned. Chapter after chapter, their book offers up changes in society that have happened so slowly and gradually that they have managed to stay under the radar for a long time even though, after a number of years, the changes are so substantial that everyone would have noticed had they occurred overnight. One of their examples is the gradually increasing percentage of Americans who stay unmarried; another is the gradually decreasing amount of time that people sleep per night. It is said that a frog that falls into hot water will jump out immediately; but if it is sitting in water that heats up gradually, then reportedly the poor animal will fail to notice the change, and allow itself to be boiled. (Don't try this at home.) In many respects, it seems, people are just like frogs: by focusing on the here and now, we fail to notice long-term trends that may be of vital importance to us.

Academic perspectives on vagueness

Symbolic logic is the science of valid argumentation. Eubulides' paradox is a case of argumentation gone mad, so if we want to understand what is wrong with sorites, symbolic logic is a natural place to look. The basic principles of classical logic, the oldest and most well-established variant of symbolic logic, will be sketched in Chapter 8. Classical logic uses what is known as the law of excluded middle, according to which every statement is either true or false, a principle sometimes associated with the name of George Boole. Eubulides' paradox appears to challenge this law, which makes it hard to deal with such subtleties as something being 'nearly' or 'very' true. Some logicians have argued that this is a flaw in classical logic, and constructed alternative logical systems, where true and false are no longer the only options. In the most drastic deviations, logicans have designed systems where truth and falsity are no longer absolute, but gradable: just like one person can be taller than another, one claim may be a little 'truer' than another. The middle part of the book will be devoted partly to these and other non-standard logics.

The second main ingredient of this book is *linguistics*: the science of language. Linguists have become a modest tribe. Long gone are the schoolmasterly days when they told the rest of us how to speak or write: linguistics has become an emphatically empirical enterprise. Because observation of human behaviour lies at the heart of this method, it is ordinary speakers and writers who determine collectively whether linguists are right—not the other way around. The present book will not limit itself to this empirical flavour of linguistics. This is partly because I will be asking *why* we speak or write the way we do, a question that will steer us towards psychology and game theory. It is also because I feel a certain affinity with philosophers such as Bertrand Russell, who did not shy away from criticizing the way language works: if people speak opaquely or misleadingly, then surely this is something that scientists should note.

Artificial intelligence (AI) is that part of computing science where programs are built that mimic human abilities, including reasoning and speaking, which are infested by vagueness. AI involves the construction of working models (e.g. computer programs) of a human ability. The same constructive spirit has moved some archeologists, for example: to understand what it must have meant to build a particular ancient building, they do not stop at analysing its structure: they proceed to construct a similar building, using original materials; reputedly, this approach led to the discovery that rice was used in the construction of the Great Wall of China. One of the main challenges for AI is to build programs that communicate effectively with people. We will have ample opportunity to investigate this challenge, culminating in the question when and why it is helpful to communicate vaguely.

This challenge brings us to the last of the disciplines from which we will borrow substantially, namely *game theory*. Symbolic logic is well equipped to answer questions about meaning and inference, but it is less obvious that it has something to say about social interaction and the ways in which linguistic utterances form a part of it. For this reason, we shall look at game theory now and then—the study of rational social interaction—for inspiration. Game theory and decision theory will come to our aid when we try to reason about the usefulness of an utterance and the merits of vagueness.

Things to remember

◆ This book is concerned with the role of *vagueness* in communication. By definition, vague expressions admit *borderline* cases. Expressions that are not vague will be called *crisp*.

◆ The main questions posed by this book are: How is the meaning of vague expressions best understood? Why does their meaning vary so much from one context to another? Why is vagueness so prevalent in human communication? And under what circumstances is vagueness preferable over crispness? We shall also be alert to any differences that might exist between different types of vagueness.

◆ Ample anecdotal evidence suggests that vague concepts are often treated as if they were crisp, without clarification of the thresholds that were employed to make them crisp. The result is a quasi-precision that can end up misinforming. Discussions of this phenomenon can be found in recent books by Blastland and Dilnot, by Paulos, and by Dawkins.

◆ One important manifestation of the difficulty people experience when reasoning with vague concepts is the *sorites* paradox, which will be discussed extensively in Chapters 7–9. This paradox has considerable practical relevance, particularly where people are confronted with gradual changes. If, for example, an increase in air pollution or population density happens slowly enough then it can easily go unnoticed for a long time, until the cumulation of small increases results in disaster.

Part I

Vagueness, Where One Least Expects It

2

Sex and Similarity
ON THE FICTION OF SPECIES

We have seen that vagueness affects many of the qualities that we ascribe to people and things. But surely, it does not affect the central concepts that we use to classify the world around us? It is one thing to say that words such as 'tall' lack crisp boundaries, but it would be something else to make the same claim about words such as 'tiger' and 'telephone'. Roughly speaking, the distinction appears to coincide with the distinction between English adjectives and nouns: adjectives (such as 'tall') denote subtly varying qualities and may therefore be vague, but nouns (such as 'tiger') denote natural classes of things and are therefore crisp, one might hope.

To show that this hope is ill-founded, we will now focus on a central building block of our thinking: the concept of a *species*. Expressions such as 'common chimpanzee' denote a species, and so does *Homo sapiens* (i.e. man). Species are the bedrocks of biology, more stable than more inclusive biological groupings such as genus, class, and order, which are often revised in the light of new evidence, and also stabler than less inclusive ones, which subdivide the species. All of these other groupings are more difficult to justify scientifically; equally, none are as entrenched in everyday conversation as the names of species. One might therefore expect that species-denoting terms have well-defined, crisp borderlines.

Biologists started thinking systematically about these matters only fairly recently. Even the famous Linnaeus, who put biologists' thinking about species on a solid footing around the year 1750, appears to have believed that there is not much of a problem here. All species are different from each other, aren't they, so it is just a matter of working hard to discover what the differences are. But in later years, particularly when it became plausible that species had developed gradually over time (rather than being created collectively in one mighty gesture), biologists realized that it would be good to have a firm principle for deciding whether two animals belonged to the same species. Around 1940 the work of people such as Theodosius Dobzhansky and Ernst Mayr had started to culminate in something approaching consensus. In what follows, let us sketch what this near-consensus amounts to.

What is a species?

Simply put, a species was defined to be a group of organisms whose members interbreed with each other. The idea is essentially this: when looking for boundaries between species, not just any boundary will do: we want a species to consist of organisms that are reasonably similar to each other in important respects. (What else is the point of grouping them together?) But how similar exactly? How do we prevent a situation in which each individual biologist has his own idiosyncratic understanding of what it takes to be a tiger? To solve this puzzle, biologists came up with the elegant idea of invoking *interbreeding* as a criterion: if two animals can interbreed then they belong to the same species. The crisp concept of interbreeding gives sharpness to what would otherwise threaten to be a fuzzy boundary. This is done on the plausible assumption that if two animals are similar enough to interbreed then their offspring must once again be similar to the parents. (Nature could conceivably have worked differently, by making offspring as different from their parents as possible, but this is clearly not what we see around us.) But is interbreeding a

20

well-defined concept, and is it as crisp as one would like it to be? Uncomfortable questions may be asked. For example:

- The notion of interbreeding applies only to organisms that reproduce sexually, so the standard definition does not apply to other species. For this reason, let us leave single-celled organisms and other celibate life forms aside and concentrate on the rest of us.
- Horses and donkeys (and reputedly even lions and tigers) can produce offspring together, but none that is fertile, so their mating does not have long-term effects. Presumably, fertility of offspring should be taken into account in the definition of a species.
- Some types of animals that do not mate under normal circumstances can be induced to interbreed with a little encouragement. Should these animals be counted as interbreeding with each other or not?
- The notion of interbreeding should disregard such trifles as age and gender. If interbreeding was the sole criterion then only members of opposite sexes could belong to the same species; consequently a species could have two members at most! This problem might be repaired, for example, by stipulating that if two individuals can each interbreed with a third individual, then the first two must also belong to the same species.
- Chihuahuas and Great Danes do not produce puppies together, because of their difference in size (although a lack of inclination might also play a role). Presumably this does not justify regarding them as different species.
- Some animals fail to interbreed because they are geographically apart from each other. If only that waterfall or mountain range did not exist, they would happily mate. (I am reminded of high school trips when desperate teachers tried to keep us in our own sleeping quarters.) It seems reasonable to disregard geographical and temporal separation, and to focus on whether two individuals *could* interbreed, if given a reasonable chance.

Taking complications of this kind into account, a species is usually thought to be something like the following:

> *Species*: a maximally large group of animals, such that healthy young specimens of the right age and sex are able in principle to produce fertile offspring under favourable circumstances that occur naturally.

One might think that all the obvious wrinkles in the notion of a species have now been ironed out. Enter the Ensatina salamander.

The Ensatina salamander

Ensatina salamanders live along the hilly edge of California's Central Valley. They tend to avoid the centre of the valley, presumably because of the heat there. Ensatina come in half a dozen different forms, which are usually viewed as its subspecies. Two of these, *Ensatina eschscholtzii* and *Ensatina klauberi* live geographically close but they do not interbreed with each other. Yet a reasonable case can nevertheless be made that they belong to the same species: *eschscholtzii* does mate with a third subspecies living just north of it; these mate with a fourth subspecies, these mate with a fifth; and these, in turn, mate with our old friends *Ensatina klauberi*. The reasons why two salamanders can or cannot get fertile offspring are buried somewhere in their physiology. For us, these reasons do not matter: we are only interested in the facts on the ground. These facts are depicted schematically here; we abbreviate the names of all but the two most crucial types of salamanders using the first letter of their names (see Fig. 2). Our scheme uses a broad brush: a very precise account would talk about *individual* animals; it is convenient, however, to simplify a little, by grouping all the salamanders in a particular subspecies (e.g. all the members of *Ensatina eschscholtzii*) together, pretending that they are all alike. (Having read the next page, you won't have difficulty reconstructing the story in a more precise style, focusing on individuals.)

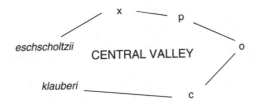

FIG. 2 Schematic diagram of interbreeding among Ensatina salamanders

The fact that *eschscholtzii* and *klauberi* live near each other allows biologists to observe their lack of interbreeding in the wild. But, logically speaking, this does not matter: their habitats might have been far apart from each other. For this reason, we can schematize even more, arranging the six kinds of Ensatina in a sequence from E_1 (this is *eschscholtzii*) to E_6 (this is *klauberi*), in such a way that each member of the sequence interbreeds with the previous one and the next one in the sequence:

$$E_1 \heartsuit E_2 \heartsuit E_3 \heartsuit E_4 \heartsuit E_5 \heartsuit E_6$$

In order not to make the creatures look more promiscuous than necessary, assume that this shows all the interbreeding that goes on: E_1 does not interbreed with $E_3, \ldots, E_6$, for example; similarly, E_2 does not interbreed with $E_4, \ldots, E_6$, and so on. If these are the facts about breeding, what does this mean for the definition of Ensatina? Do all six types of salamanders count as Ensatina?

The answer, based on the standard definition of 'species', is No! Sure enough, E_1 and E_2 form a species together, and so do E_2 and E_3. But E_1 and E_3 do not, hence they do not belong to the same species. Note, however, that the two species that we have just found ($\{E_1, E_2\}$ and $\{E_2, E_3\}$) overlap, because E_2 occurs in each of the two. Instead of being nicely separate chunks of fauna, these two species end up all intertwined (see Fig. 3). This, surely, is not how we like to think about animals.

How have biologists responded to the problem set by Ensatina and about two dozen similar cases reported in the biology literature? The dominant reaction appears to be to let common sense prevail over

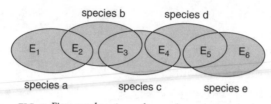

FIG. 3 Five overlapping salamander species (a)–(e)

formal definitions. The standard view is roughly thus: E_2 belongs to the same species as E_1. But E_3 belongs to the same species as E_2, therefore E_3 must belong to the same species as E_1. The argument can be repeated for E_4: it belongs to the same species as E_3, therefore it too must belong to the same species as E_1. Similarly, the argument can be repeated for E_4, for E_5, and for E_6, showing that all six types of salamander belong to one big species. This standard view, however, lumps together all the different types of salamander into what is sometimes called a *ring species*.[1] This includes E_1 and E_6, which do *not* interbreed together and which would therefore not belong to the same species, if the standard definition of 'species' had anything to do with it. Fortunately, E_1 and E_6 are still a bit similar—they are all ordinary salamanders. If there happened to be many more animals in the world, then perhaps there would exist animals E_7, E_8, and so on, each of which is a little different from the previous one, including some E_n (where n is some seriously large number) which is entirely unlike any salamander... and a biological version of the sorites paradox would arise!

How should we view the story of the Ensatina salamander? Is it indicative of an important flaw in the concept of a species, or just a little anomaly about which we should not worry? We shall use the remainder of this chapter to answer this question.

To see how the story of Ensatina highlights a much more general problem, let us follow the biologist Richard Dawkins, who has written engagingly about Ensatina, by focusing on ordinary situations, when a species develops over time. Our own wretched species will provide as good an example as any.

Suppose you were to draw up a huge list of your ancestors, always choosing a parent whose sex is opposite to your own. So if you are a woman, you list your father, paternal grandfather, and so on. Suppose you had superhuman amounts of time and patience, going back to the person who lived 250,000 generations or so ago, something like 6 million years BC. Now we need to ask some rather personal questions. Would you be able to interbreed with the first ancestor in the sequence (your mother or father, that is)? Unpalatable though the thought is, I take it that the answer is 'probably yes'. How about the second ancestor (one of your grandparents)? Well, if we disregard time and emotion then probably the answer must be yes. But at some time t in the past, so many biological differences will have accumulated between you and the ancestor in the list who lived at that time t that it would be impossible for the two of you to have fertile offspring even if you had been the fondest of contemporaries. I do not know precisely how far back in time we need to go, but 6 million years ago is almost certainly long enough ago, since this is estimated to be the time when your ancestor was also an ancestor of today's chimpanzees: probably an ape who walked on four legs. For future reference, let's call him Richard.

Recall the story of the Ensatina salamander: Ensatina is often treated as just one species of salamander. This seems sensible. What do all these tiny differences between six types of animals matter, given that each interbreeds with the next one? But suppose we applied the same logic to the human species, based on the story involving Richard? We would start reasoning that you are the same species as your parents, who are the same species as their parents, and so on. After 250,000 or so reasoning steps, we would be concluding that your ancestors 6 million years ago—walking on four legs, having no language and hardly using any tools—were human too. If this is not bad enough for your taste, then let's reverse direction: starting from Richard, we keep choosing a child instead of a parent: first we choose one of Richard's children, then one of her children, and so on. By making the right choices often enough, we arrive at a chimpanzee now living in the London Zoo. Using the same reasoning as before, we conclude that this chimp is of the same species as Richard, who is of the same species as you. Chimps would be human, in other words.

Let us reflect on the situation. First, the reasoning is not limited in any way to chimps and people: if we follow current biological orthodoxy and accept that all today's life evolved from one and the same source, then all living beings are cousins of each other. Therefore, the reasoning above can be applied with equal force to, say, trouts and tigers: the only difference is that the required number of reasoning steps would be larger in that case, because the latest common ancestor shared by trouts and tigers is probably almost half a billion years ago. So if you buy into the standard way of reasoning about species then all living beings are one and the same species. This conclusion may be morally pleasing, but it makes a mockery of our attempts to classify the world.

Secondly, the reasoning seems far-fetched primarily because we are not acquainted with all the intermediate types of animal that link you to the ape called Richard, and Richard with today's chimpanzee. It is not only that most of them are dead, for the reasoning could also be applied to fossils; but far too few fossils have been found to perform this reasoning on the basis of them: we simply don't have enough of them to be able to arrange them in the kind of 'n interbreeds with $n + 1$' sequence that gave rise to the argument. This is where Ensatina was different, since so many intermediate forms are alive and well, geographically close enough to each other to rub our noses into their breach of biological principles.

The story of the Ensatina salamander reached a wide audience through Dawkins's book *The Ancestor's Tale*, a splendid account of evolution, organized as a journey back in time, all the way to the beginning of life. It will be useful to examine Dawkins's conclusions in some detail. Reflecting on people's tendency to think in terms of discrete categories, he rejects what he calls the tyranny of the discontinuous mind.[2] Following Ernst Mayr, he traces this 'tyranny' back to Plato, who famously believed that words reflect a fixed and unchangeable pattern that is independent of human thought.[3] In Plato's philosophy, these patterns include not only such basic concepts as a circle—for which the Platonist position has undeniable appeal—but also such words as 'man', 'dog', 'table', and so on. Dawkins rightly has little time for this outdated manner of thinking. But if Platonism is the wrong way of thinking about biology, then what is the correct way of thinking about, say, species? Dawkins decides to keep using species terms, while taking them with a pinch of salt: 'Let us use names as if they really reflected a discontinuous reality, but let's privately remember that, at least in the world of evolution, it is no more than a convenient fiction, a pandering to our own limitations' (Dawkins 2004: 320). Dawkins observes that our minds work in a 'discontinuous' fashion. When exactly this is acceptable, rather than a tyrannical aberration, is a question that he leaves aside. In the case of Ensatina, I take him to accept that this is just one species, even though this contradicts the standard definition which dictates that there are five (i.e. one consisting of E_1 and E_2, one consisting of E_2 and E_3, and so on). If this approach were applied rigorously—which is something Dawkins would presumably advise against—then the total number of species that ever existed in the world would be just one! In the presence of inconsistency (i.e. between entrenched species terms on the one hand and the definition of 'species' on the other) it is easy to stumble.

Lessons learned

What exactly is it about the biological examples discussed in this chapter that causes them to be problematic? To answer this, let us return to the Ensatina salamander. Suppose you are a biologist in the year 3000, studying Ensatina. Progress being what it is, it seems likely that several of the six types of Ensatina salamander will be extinct. Let us assume that the remaining ones will not have changed very much. Suppose, concretely, that E_2, E_4, and E_6 have gone extinct and that, somehow, all traces of them (including the writings of biologists) have disappeared off the face of the earth. The schematic picture that we drew earlier in this chapter can now be simplified substantially. All you will find is three species of salamanders, E_1, E_3, E_5, none of which interbreed with each other.

In the new situation, vagueness has disappeared. In our own day and age, Ensatina salamanders caused us trouble because not enough of their subspecies had the decency to become extinct. Biological taxonomy is often unproblematic in practice precisely because all manner of intermediate cases have died out, leaving us mostly with species all of which are neatly distinct from each other. Ensatina is an exception that opens our eyes to how things might have been.

There is a much more general lesson here, namely that the suitability of a concept depends on the abstract structure of the reality described by it: in the world as it is, 'Ensatina' is a problematic concept, but if things develop as we imagined then it will be unproblematic in the year 3000. The key difference between the two situations is this. In the year 3000, one can state as a general rule for the salamanders of interest that if one interbreeds with a second and the second with a third, then the first interbreeds with the third as well. In a mathematician's jargon, the relation 'interbreeds with' will be transitive.[4] In the year 2000, the relation is not transitive. In the chapters about logic, we shall see that non-transitive relations of this kind lie at the root of much evil.

The rather formal considerations that were examined in this chapter are not the only ones driving a field biologist's decisions. Interbreeding, after all, is used as a criterion for specieshood only because it tends to lead to a grouping of animals that are similar in important respects. Furthermore, information about an animal's breeding pattern can sometimes be hard to come by, making the standard definition difficult to apply. In practice, therefore, biologists let genetic, organic, and behavioural similarities play a role alongside interbreeding. None of this, however, invalidates the claim that species-denoting terms are vague. Quite the contrary, it complicates the situation even further, allowing different biologists to make subtly different decisions from case to case.

Before we move on to other pastures, it is important to realize that the story of this chapter is not just about biology. We have focused on the notion of a species, but other concepts that classify things into groups can be analysed along the same lines. Perhaps the case most similar to that of a species is that of a *language*, as when we say that British and American English are (or are not) the same language. We shall turn to these issues in Chapter 4, but first we shall talk about measurement.

Things to remember

◆ The classical notion of a species based on interbreeding is an attempt by biologists to give all species crisp (i.e. non-vague) and objective boundaries, while ensuring that species are held together by physiological similarities.

◆ So-called *ring species* cause problems for this classical notion of a species because interbreeding among them is not a transitive relation.

◆ The classical concept of a species is not consistent with the way in which words such as 'chimpanzee' or 'tiger' are actually used. Rather than a properly shaped tree of life, the criterion of interbreeding leads to a useless gamut of mutually overlapping species.

◆ The alternative of joining all Ensatina salamanders into one species is not satisfactory either. When taken to its logical consequence, this approach would cause all living things to be regarded as one species, which would rob the concept of its usefulness.

◆ Terms such as 'tiger' are nevertheless useful because all animals similar to tigers have become extinct. These terms lose much of their appeal when they are applied across a period of time during which important changes occur to the organisms of interest.

♦ Similar problems affect other concepts by which we classify the world. An example is the notion of a *language*, as defined by the criterion of mutual understandability (according to which two dialects are part of the same language if their speakers can understand each other). Languages are vaguely defined in much the same way as species.

3

Measurements that Matter

The Scots have a sport called 'Munro bagging'. A Munro is a mountain whose summit reaches at least 3,000 feet (914.4 metres) above sea-level. The idea of the sport is to climb as many Munros as you can. As far as I can ascertain, there is nothing intrinsically important about the threshold of 3,000 feet. It's just that one Sir Hugh Munro, back in 1891, used it as his criterion when he published his famous tables of Scottish mountains, 'which thus provide the outdoor fraternity with endless challenges', according to the *Collins Encyclopaedia of Scotland*. Some critics have questioned the legitimacy of Munro bagging as a serious sport—one of them likened it impertinently to kissing all the lamp-posts in Edinburgh's Princes Street—and specifically the arbitrariness of the 3,000-foot boundary. Without taking sides in so momentous a debate, one can see what these critics are driving at.

Underlying every quantitative assessment—that something is large or small, high or low—there lies a measurement of some sort. To see what goes on when things are measured, let us take a brief look at a range of phenomena, asking how the relevant phenomena are measured. Where significant boundaries are in use (like the 3,000-foot boundary in Munro bagging), we shall ask what motivates them.

A short history of the metre

Eindhoven's University of Technology proudly displays a slogan by the Dutch physicist Kamerlingh Onnes at one of its buildings: 'Door meten tot weten' ('Through measurement to knowledge'). Measurement is essential to the acquisition of knowledge, but measurement is not always easy. Consider, for example, that most fundamental of measurements: the measurement of physical distances, as when we use feet and yards, for example, or metres. Once upon a time, a foot must have been thought of as the size of, well, a human foot, without worrying whose foot exactly. (It is said that carpet salesmen tended to have small feet!) These days, we are no longer satisfied with this level of imprecision, of course, and concepts such as the foot and the metre have long been standardized—at least to an extent. Since metres are the international currency in most areas of science and engineering, let us briefly look at the history of the metre.

How do you define a metre, a yard, or a verst in an objective way, so anyone can check whether their measurement instruments are accurate?[1] It does not matter what we decide, say, a metre to be, as long as everyone understands the concept in the same way. The natural idea is to relate the unit in question to some known and invariable thing that is accessible to anyone: a *bedrock* as we shall say. An early and interesting example is Huygens's idea, way back in 1673, to base measurements on the distance travelled by the swing of a clock in a second. Others chose the distance travelled by a falling object in one second as their bedrock. Many of these early attempts at finding a natural bedrock, however, suffered from the fact that gravity varies across the surface of the Earth, because the Earth is not perfectly round. Accordingly, none of them managed to gain general acceptance.

As accurate measurement became more and more important, the French government declared, in 1790, the ideal of establishing a system of measurements suitable 'pour tous les temps, pour tous les peuples' ('for all times, for all peoples'). To arrive at such a system, the French government initiated a research project whose ultimate bedrock was mother Earth itself: a metre was equated to 1/40,000,000 of the length of the meridian that runs from the North to the South Pole, exactly through the Panthéon, a building in Paris which had started its career as a church, but had graduated to a 'temple of reason' during the French Revolution. Naturally, it took scientists a while before they figured out the implications of the Parisian definition. In the end, they produced a bar of platinum that they argued to have just the right size. From this moment onwards, this bar became their de facto bedrock.

It was of the essence, of course, that the size of the bar was kept as constant as possible. For this reason, it was to be kept always at the same temperature, at zero degrees Celsius, the melting point of ice. Platinum was chosen so as to minimize corrosion and any remaining fluctuations caused by temperature. Unimaginable care was taken to give the bar an optimal shape, mainly to make sure that it bent as little as possible between its two supports. A lot of work was done in subsequent decades to make this standard metre bar as stable as possible, and to allow it to be copied for the benefit of people unfortunate enough to live far away from Paris. But tests in the twentieth century indicated a lack of precision, in measurements based on the Parisian metre, of about 0.00005 mm (0.05 μm).

A measurement error of 0.00005 mm may sound like a trifle, but in microscopy, for example, it isn't. Even for the construction of a humble watch, one needs machinery that produces tiny nuts and bolts whose dimensions match their specification very closely, or else the two won't fit together. Astronomical and medical instruments make even higher demands. The Parisian approach had been highly successful (having been in use for about a century) but new technologies demanded greater precision and it was difficult to see how the error of 0.00005 mm could be reduced much further. Perhaps surprisingly, the main problem was posed by the difficulties involved in deciding where the bar starts and ends. The standard method became to use a slightly longer bar, on which the start and end of the metre were marked with tiny incisions.

But even tiny incisions have a width, causing vagueness. All this meant that researchers kept looking for a bedrock more stable and clearly defined than the Earth or any object on it.

One improvement, proposed around 1900, was made by choosing the wavelength of a particular kind of light (multiplied by some suitably large number, of course, to arrive at a distance close to the Parisian metre) as the new bedrock. Not only did this lead to a considerable reduction in measurement errors, it was more elegant as well. In comparison with the properties of light, the size of the Earth is a mere accident! Perhaps most important of all, the new definition had practical advantages. In the old situation, in which a platinum bar in Paris called the shots, a producer of measurement tools living in Boston, say, would have had to travel to Paris, unless he was confident enough to base himself on American equivalents. (If he was lucky, the Bostonian metre bar was a direct copy of the Parisian one. More likely, it was an error-increasing copy of a copy.) In the new situation, where *light* is the bedrock, it doesn't matter whether you are in Paris or Boston: the distance is always the same. The basic idea of using wavelengths as the bedrock of measurement was refined in subsequent years, for example by using wavelengths emitted by Cadmium (around 1927) and Krypton 86 (around 1960). Interestingly, some recent definitions combine the idea of using light as the bedrock with Huygens's old idea of using time: the Metre Convention of 1983 introduced a new definition whereby a metre was equated with the distance travelled by light through a vacuum in $1/299{,}792{,}458$ of a second.

It is safe to assume that the search for stabler and stabler definitions will go on, and that new definitions will make increasingly sophisticated use of physics. Some sobering conclusions are worth drawing. First, some of the basic notions underlying our everyday activities (as well as science) are subject to change and reinterpretation: a metre is a different beast now from what it was in the year 1700, 1800, or 1900. Secondly, these notions are defined in increasingly sophisticated ways, which require the work of skilled physicists, leaving the rest of us with mere approximations. (Pretty good approximations, but that's a different matter.) So even though measurement terms are used by everyone, their definitions have effectively been outsourced to science. Distance is one of the easiest things to measure. Weight, speed, temperature, barometric pressure, you name it: their measurement involves all the same problems as distance, and then some additional ones as well. In each of these cases, scientists have had to work hard to find definitions that are at least reasonably precise.[2]

But, crucially, even the most sophisticated definitions have some residual imprecision, and it is unclear how this imprecision can ever be got rid of completely. Suppose one uses a definition based on the speed of light, for example. Although it is difficult to know with absolute certainty that the speed of light is always exactly the same, let us grant that this is the case. This means that, in theory, the definition of a metre will now be precise. But *in practice*, the definition can only be applied to the measurement of an object by getting one's hands dirty, that is, by measuring the time that it takes for a beam of light to travel the length of the object. This measurement will involve procedures that are bound to involve error, and this error will reintroduce vagueness. Moreover, the definition refers to the speed of light that travels through a vacuum, and the construction of an absolute vacuum is difficult to achieve in practice. Measurement reports involving metres will probably always remain vague: if a distance is close enough to 1 metre then some measurements may estimate it as having a size just below 1 metre while others may estimate it as having a size of just above 1 metre, for example. If this happens, it is unclear whether the distance is at least a metre or not: it will be the kind of boundary case that is the hallmark of vagueness.

In cases like this, paradox is never far away. Suppose you find a platinum bar in a forgotten attic. To your delight, you measure it as having a length of exactly 1 metre. During the night, a thief shaves off one end of the bar, just a tiny slice of one thousandth of a μ. The next morning you measure the bar again, concerned that anyone might have been messing with your treasure. But surely no measurement tool is precise enough to allow you to notice the loss of such a thin slice with any kind of confidence... The following night the thief returns and removes another slice, too thin for you to notice, and so on. If all you and your heirs will ever do is compare the size of yesterday's bar with today's bar—which would not be very smart, of course—then no one will ever notice the difference. Paradoxes of this kind will be discussed at length in the chapters on logic.

Measurements are often associated with words. Words often lack a generally agreed meaning: if you look up a word in a dictionary you will often find it to be defined in a number of different ways. Were you to ask the native speakers of the language, or to verify how they use the word, you would discover even greater differences. Sometimes this lack of unanimity is of limited interest. But when a word or concept matters greatly to a community, then lively disagreements may arise. The upshot of these discussions is often a ceasefire, during which a particular understanding of the word finds wide acceptance. After a while, new hostilities may break out, until a new (and perhaps equally short-lived) consensus is reached. Let us examine a few of these measurements that matter, to see what questions have come up and how they have been dealt with.

Obesity

Let us start with a phenomenon not dissimilar to the previous one, which involves the size of our body. Obesity is regularly in the news these days. Politicians and newspapers declare that 'one fifth of British school-aged children are obese', 'In the USA, adult obesity has doubled between 1980 and 2002', or variants on the same theme. As we noted in the Introduction, it is striking that such claims are made as if we all knew what obesity meant, and that they presuppose that to be obese or not is an all-or-nothing affair. Let us see why the authors of these claims can get away with them.

The currently prevalent perspective on such notions as 'overweight' and 'obesity' is profoundly *medical* in outlook: to be obese means, roughly, being too tubby for your own good.[3] It's interesting to wonder whether this is a novelty, or part of a much longer tradition. Why, for example, did Rubens paint his little cherubs as if they were all overweight? In the dark days before McDonald's, overweight was a sign of wealth rather than poverty. In the seventeenth century, when Rubens painted, diseases such as the plague were still going strong, and the fact that the cherubs were overweight meant that, at least, the little dears were not about to give up the ghost. It may be worth noting that, even now, the beauty ideal of the fashion show—populated by the 'super-thinnies'—is not the only one going. It has been claimed that Aids has a Rubensian effect in some sub-Saharan countries in our own time. In desperate times, to look thin is to look ill.

Be this as it may, the concepts of overweight and obesity that are used by Western politicians and news media are even more explicitly medical. Whatever metrics of obesity are used, their motivation comes not from aesthetic judgement—more than a few of today's beauty queens would count as officially *underweight*—but from medical research. The basic idea is simple: 'If weighing over 200 kilos is likely to make you ill then that's bad. We'll call you obese.'

Gradually, people started to suspect that it's healthier not to be too heavy. This may have started with anecdotal evidence, but at some point figures became available showing that heavy people are more likely to suffer diseases such as type-2 diabetes. Then in 1983, analysable data became available: the insurance company Metropolitan Life published its tables, revealing a strong statistical correlation between obesity and mortality: the heavier you are in relation to your body frame and age, the shorter your life expectancy. Soon the question was no longer whether there is a link between obesity and health, but what the precise nature and causes of this link are. Defining obesity became a doctor's game... and a statistician's.

The game goes roughly as follows. You want to find a definition of obesity that optimally explains the bad health of the people who are obese. You start by focusing on one particular way of measuring the health of a population, for example using mortality, or average incidence of type-2 diabetes. Now, you make a wild guess: you guess by choosing a *dimension* that might be suitable for defining obesity. For example, you could start focusing on body weight. Then you start gathering data about lots of people: their body weight at 40 years of age, for example, and the age at which they die. Then you analyse your data, asking how strongly the two variables are related. Presumably, for example, people's chances of reaching 70 are not equal for all body weights: this chance will tend to be highest around some 'normal' weight, and diminish gradually for people further removed from this weight. This is the easy part.

The difficult part is to *vary* the decisions that you have tentatively taken, and see how this variation affects the trends. In all likelihood, the naive procedure described above will not give you a very strong correlation. Pondering over your data, it may occur to you that plenty of people who are very heavy are also very healthy, for example when they are tall and muscular. If this is what you think, then the natural move to make is to measure degree of obesity in more sophisticated ways, by going beyond a simple weight measurement. Suppose your data contain two people of equal weight, one of whom is much taller than the other; then you may want to say that the shorter of the two is the more obese. Having redefined obesity, you now analyse your data once again, wondering whether the correlation between newfangled obesity and ill-health is stronger than that between oldfangled obesity and ill-health.

In theory, we can continue this game till kingdom come: there is no way of knowing that you've found the 'best' definition of obesity. Practical considerations can play a big role. Excess body fat, for example, makes a person less healthy, because fat causes your heart to have to work harder. Yet this does not mean that body fat should necessarily form a part of the definition of obesity, for whereas it's easy enough to measure someone's body weight, it is much harder to measure their body fat. Practical concerns of this kind have caused epidemiologists to settle for a measurement that is theoretically less than optimal, but practically feasible, namely the body mass index (BMI), proposed by the Belgian statistician Adolphe Quetelet. Let's see to what extent it is possible to understand why the BMI formula was chosen. A person's BMI is a function of his weight and height, as follows:

$$\text{BMI} = \text{weight}/\text{height}^2$$

Let's suppose that the ideal BMI is about 20. To see what this BMI dimension amounts to, let me introduce three people to you: Mr A is very short, with a height of 1.00 m, Mr B is very tall at 2.00 m, and Mr C is a giant of 4.00 m. (We've doubled the size of Mr A twice.) There is nothing intrinsically good or bad about being short or tall, so it's fair to ask what the ideal weight of the three people would be, given their height. The formula for BMI, together with the claim that a BMI of 20 is best, implies that weight/height^2 = 20. Filling in the heights of our three friends, with w for weight, this means

Mr A: $w/1.00^2 = 20$, hence $w/1 = 20$, so $w = 20$ kg
Mr B: $w/2.00^2 = 20$, hence $w/4 = 20$, so $w = 80$ kg
Mr C: $w/4.00^2 = 20$, hence $w/16 = 20$, so $w = 320$ kg

In other words, BMI says that by doubling someone's size, we ought to quadruple his weight. More generally, multiplying someone's height by n, we ought to multiply their weight by n^2. In mathematical jargon: ideal weight grows quadratically with body height.

If people had the shape of a die (which, admittedly, is rare) then their body weight would grow much faster than what is implied by the BMI formula: if you double the dimensions of a cube of height n, its volume is not multiplied by n^2, but by n^3: cubically, as mathematicians say. Yet, BMI embodies the theory that body weight increases quadratically with height. To appreciate the idea, let us return to Messrs A, B, and C and wonder how their frames might differ if they were ideally proportioned. It's unlikely that anyone looks like two copies of A on top of each other, let alone four. It would be better if B, being twice as tall as A, is also almost twice as broad-shouldered, though not much fatter. If your arm were twisted and you had to find a quick approximation, you might say that B should be shaped as follows: first, you stack two copies of A on top of each other; then you form another stack that's a copy of the first; to stabilize the construction, you put the two next to each other. This simple construction is consistent with the theory embodied by the BMI formula. Here is how BMI is applied to adults.

Underweight: BMI < 18.5
Healthy (or normal) weight: BMI between 18.5 and 24.9
Overweight: BMI between 25 and 29.9
Obese: BMI between 30 and 39.9
Morbidly (or severely) obese: BMI > 40

The cut-off points (18.5 and so on) are of course arbitrary to a large extent: there is no decisive medical evidence against increasing or decreasing each of them by a point or so. It is therefore striking how near-unanimously these cut-off values are presented throughout the literature following their publication in 2000. Variations and refinements have been proposed, but these have tended to be small enough to be negligible. Given the simplicity and arbitrariness involved in the definition of BMI, however, it seems plausible that the future will see some very different definitions of obesity and overweight.

Poverty

Having noted some of the problems involved in the development of useful definitions of obesity and overweight, let us move to a slightly more complex case, which inhabits the social rather than physical domain. The twin concepts of poverty and wealth have dominated the science of economics, at least until recently, when softer concepts such as quality of life began to assert themselves. We focus on the strictly financial aspect of poverty, and more particularly a person's income, leaving inherited wealth aside, for example, relevant though it may obviously be.

Social phenomena are often more difficult to pin down than physical ones, and poverty is no exception. A recurring question is whether poverty should be defined in absolute or relative terms.[4] An *absolute* definition of poverty might, for example, link the concept to what our bodies need in terms of food and shelter, analogous to the physical bedrocks that were used to pin down the metre. Understood in this way, modern societies have made great strides in combating poverty, with the notable exception of disenfranchised groups such as illegal migrants. But as long as people have written about wealth, it has been recognized that the concept of absolute poverty has limited value as a tool for understanding human behaviour. Suppose you grew up in a neighbourhood where everyone earns at least three times as much as you do. You may be well able to buy food and shelter, but if you do not earn enough to take part in the leisure activities of your friends then this may well make you feel 'poor', and your comparative poverty could exert a considerable influence on your life. You might even be tempted to evade taxes or to steal. The fact that many people resist such temptations is beside the point: it only takes a small proportion of discontents to cause mayhem, which could affect the rich as well as the poor. It is partly for selfish reasons of this kind that *relative* notions of poverty were introduced.

Winston Churchill appears to have held that social security should be no more than a safety net 'holding people just above the abyss of hunger and homelessness'. But the discussion appears to have moved on, at least in the United Kingdom. In a recent policy paper, Mr Greg Clark, a Conservative politician, argued that Churchill's view is no longer appropriate for Conservative social policy in the twenty-first century, adding that his party should take more note of the views of the well-respected newspaper columnist Polly Toynbee, who is usually associated with a very different section of the political spectrum. Toynbee was singled out because of one of her columns in the *Guardian*, where she

41

had compared a nation to a caravan that travels through a desert. Her point was that collective survival depends on the wagons sticking together: no one should pull ahead or lag behind too far. Soon the Conservative party leader David Cameron was quoting Toynbee approvingly, saying, 'To be poor is to fall too far behind what most ordinary people have in your own society.' He went on, 'Even if we are not destitute, we still experience poverty if we cannot afford things that society regards as essential.'

This remarkable unanimity comes after a decade during which, in many countries, the gap between rich and poor has widened. One often-used measure compares the income earned collectively by the top 10 per cent of earners with that of the bottom 10 per cent. In 1994 in the USA, the former group earned 10.6 times as much as the second, but by 2005 this multiple had risen to 11.2. The widening of the gap appears to be largely the result of the good fortunes of the very rich. Some indications that the wagons at the front of the caravan are indeed pulling ahead, at least in the United Kingdom, are provided by the *Sunday Times* Rich List, which reported in 2007 that the wealth of the richest 1,000 people in Britain has more than trebled since 1997, while the number of billionaires born, living, or working in the UK had trebled between 2003 and 2007. Based on reports of this kind, Sir Ronald Cohen, one of Britain's leading businessmen, has been quoted as warning that the growing wealth gap could ultimately lead to violence. But even if it does not come to rioting, real or perceived imbalances in income can be risky: increased wage demands could follow, for example; higher wages, in their turn, could trigger price inflation, and inflation could trigger interest rate rises. Society as a whole would suffer.

Relative poverty can be defined in many different ways, but the discussion tends to be dominated by a limited number of definitions. Interestingly, poverty is often treated as if it were a crisp concept. One popular definition of povery, for example, defines the poverty line as 60 per cent of the median income. The median is the income level x such that the number of people earning *more* than x equals the number of people earning *less* than *x*. This makes the median a bit like the average, but it doesn't care so much about the extremes: a mere millionaire counts for as much as a multibillionaire, for example. If you're under this poverty line you're poor; otherwise you're not. Needless to say, there is more than a little arbitrariness here. Why 60 per cent, for example? The fact that the definition is based on the median is worth noting, because this means that 'fat cat' salaries are irrelevant, since company directors earn well above the median anyway. According to this school of thought, it is the people at the bottom of the ladder who are the problem, not the ones at the top: it's the wagons at the tail end of the caravan that deserve politicians' attention.

When the concept of poverty is applied to groups of people rather than individuals, relative notions of poverty remain relevant. When one nation, or other identifiable group of people, is perceived as much wealthier or more powerful than another, envy is never far away. Or consider present-day politics: it would be a gross simplification to depict the kind of terrorism that is, in the first decade of the twenty-first century, associated with some forms of Islamic fundamentalism, as exclusively caused by poverty, but there can be little doubt that relative poverty plays a role: poverty relative to other countries, for example, and poverty relative to a (relatively!) wealthier past.

An important issue emerging from this discussion is how measurements of poverty need to take the *context* of someone's economic situation into account: to judge whether someone is poor, you have to take into account not only the money available to them, but the money available to other people. Taking the context of a phenomenon into account, however, is difficult, because you need to find out what the *relevant context* is. Suppose you are assessing the economic situation of the Basques in the north of Spain, for example, to understand the political situation in that part of the world. To assess whether the people in question qualify as poor, should one compare their economic situation with that of other people in the industrial north of Spain, or would it be better to look at the whole of Spain? But given that Basques live in

France as well as in Spain, both countries may have to be considered, and so on. There are many possible contexts, each of which may have some relevance to whether the population in question might feel left behind.

This contextual aspect of poverty is an interesting issue more generally. It's as if you're talking about a hilly landscape, with hills and valleys. Whether something is a valley does not depend on its absolute altitude, but on the altitudes surrounding it. (I know a hill in the Netherlands that would count as an abyss if it was located in the Himalayan Kingdom of Nepal.) The theory of relative poverty regards financial resources in the same light. Later, when discussing the linguistics and logic of vagueness, we shall see that vagueness and context are closely linked, and that many of the issues discussed in connection with poverty will reappear there.

Enough about poverty. Socially useful concepts of wealth and poverty may not be easy to formulate, but at least these notions are grounded in something we understand rather well, namely money. Let us see what happens when we try to measure something where no such grounding is available.

Intelligence

Authors in the Romantic era liked to theorize about the concept of *genius*: does genius differ from *talent* merely as a matter of degree, or are they two qualitatively different concepts? In our democratic times, we have become more interested in concepts that are applicable to all of us, and particularly where they can be made objectively measurable. Wouldn't it be great if a person's capacity for learning, job performance, and general success in life could be measured?

The question of how to define intelligence has become one of the most hotly debated areas of psychology. The focus of the debate is usually a test that claims to measure a person's cognitive capacities. At the heart of the test is a set of puzzles for whose solution you get a fixed amount of time; the more of the puzzles you solve, the higher your IQ. Since the primary aim of the test was to predict potential success rather than present performance, the test typically involves a comparison between people of the same age. The earliest of these tests, developed between 1910 and 1920, were designed for children, and expressed as a so-called intelligence quotient (IQ) in a way that makes use of their intellectual growth through time:

$$IQ = 100 \times \frac{mental\ age}{chronological\ age}$$

According to this formula, a child of 10 performing like a 14-year-old has an IQ of $100 \times (14/10) = 140$. This definition is, of course, problematic in the case of adults and this has triggered various refinements whose details need not concern us here. Basically, a modern IQ test aims to tell you what percentage of people in your age group are smarter than you. An IQ of 100 means that about 50 per cent of people are smarter; for an IQ of 120 this is roughly 10 per cent, and for 150 about 0.1 per cent.

To accept a particular IQ test is to endorse a particular concept of intelligence. It is, after all, not obvious what kinds of puzzles these tests should consist of, and in what proportions. It is worth comparing the situation with our discussion of obesity assessments. Few people will question whether mortality and disease are worth taking into account when assessing whether someone counts as obese. But each part of an IQ test is open to precisely such questioning. It would be different if IQ tests focused on tasks that were essential for survival or procreation—such as the collection of food or the invention of chat-up lines—but they usually are not. They are, in an important sense, arbitrary, and this makes them sensitive to criticism, as their history confirms.

Fierce debate broke out, in the 1960s and 1970s, when some intelligence tests had African-Americans score lower than other groups in the United States.[5] Some concluded that there must be something wrong with African-Americans, perhaps genetically, or as a result of racial discrimination and the legacy of slavery. Others argued that there was something wrong with the tests themselves, arguing that they were phrased in a kind of language that put speakers of black vernacular English at a disadvantage. Changes were made to the tests—and the gaps between racial groups appeared to narrow—but new controversies were never far away. It is fair to say that this is still the situation today, and yet IQ tests are routinely used by more than a few institutions, presumably because a flawed measurement instrument is thought to be better than no instrument at all. The downsides of this attitude are obvious: some people who should be offered a job, for example, end up disenfranchised because their potential is underestimated. Their would-be employer loses out as well.

Perhaps the debate about racial and social issues is premature, given that it is unclear what IQ tests measure. Intelligence is an ambitious thing to measure, partly because people's mental behaviour is so multifaceted: if Mary is quicker in maths (or humour, or music) while Harry is better in languages (or spacial orientation, or empathy), then how should the two be compared? How should the various dimensions along which they differ be weighed? Analogous to the role of medical science in relation to obesity, the question could *in theory* be settled by statistically investigating the impact of each factor on a person's success in life, but IQ tests have never been particularly strong in this regard. Recent research suggests, for example, that above 120, IQ is almost worthless as a predictor of success, lagging far behind other factors, such as one's parents' income, and some people argue that IQ tests are an instrument of repression, designed to keep the masses in their place.

Perhaps it was always misguided to want to measure such an elusive concept as intelligence, particularly in the absence of a generally accepted theory of what intelligence is: if you do not know what it is that you're trying to measure then your test is ultimately built on sand. This sceptical view appears to be gaining ground in recent years, particularly after it became clear that the figures reported for IQ tests have been rising rapidly throughout the twentieth century; it is difficult to believe that our species has become rapidly more intelligent over such a short period of time. This so-called Flynn effect (Dickens and Flynn 2006) suggests that the tests should stick rigorously to comparisons within one generation, which would serve to call further into question what it is that these tests measure in the first place.

DIALOGUE INTERMEZZO
AFTER THE JOB INTERVIEWS

Stephen, Anita, and Nigel are sitting around the table, where they have just interviewed a series of candidates for a lecturing position at the University of Poppleton. Three of the candidates impressed them with their abilities. For reasons of data protection, we call them Tea Ching, Ad Min, and R. E. Search. A choice between these three needs to be made, and it is getting late in the day.[6]

STEPHEN: *It is good to have three such outstanding candidates for the job, but we have to make a choice. (Walks over to the whiteboard and draws a diagram.) I know I'm simplifying, but this is the gist of it. A lecturer does research, teaching and administrative duties, right? In terms of their research, Search is best, Ching comes next, and Min comes last. In terms of their teaching, let's see... Ching is best, Min comes next, and Search last. Finally, we have administrative skills, where the ranking is Min, Search, Ching.* (All compare the diagram with the paperwork in front of them. The diagram appears to be correct.)

RESEARCH	TEACHING	ADMIN
S	C	M
C	M	S
M	S	C

47

ANITA: *But your diagram doesn't help us much, does it? It just shows why it's difficult to rank the candidates.*

NIGEL: *What do you mean?*

ANITA: *Well, none of the candidates beats the others on all three criteria. So the best we can hope for is someone who beats the others on two of our criteria, right?*

NIGEL: *I can go along with that.*

ANITA: *Now look what happens: Min beats Search in terms of teaching and admin; Search beats Ching on research and admin. You might think that these two things imply that Min beats Ching, but watch out: Ching may trail Min in terms of Admin, but she is ahead of him in terms of the two other criteria.*

NIGEL: *Blimey, it's a cycle! The first candidate beats the second, the second beats the third, but the third beats the first!*

ANITA: *That's right. This is deadlock!* (All stare at the whiteboard.) *They're all equally good!*

NIGEL: *That's right. They're all equally good. I suddenly understand why it can happen that one football team beats a second team, the second beats a third, and yet the third beats the first. It can happen even if every team always plays equally well...*

STEPHEN: *Hmm.*

NIGEL: *Stephen says 'hmm'.*

ANITA: *Steve, why do you say 'hmm'? Is it the football?*

STEPHEN: *No, Nigel's spot on about the football. The problem is... I just got a text message on my mobile.*

ANITA: *And?*

STEPHEN: *It looks as if Min will withdraw. Another job offer. Very prestigious...*

NIGEL: *No problem. This makes our problem easier. We choose Search or Ching and go home.*

STEPHEN: *I'm not so sure about that... Look at the diagram again.* (Walks to the whiteboard and replaces Min's name by dashes.) *Do you see what happens?*

MEASUREMENTS THAT MATTER

RESEARCH	TEACHING	ADMIN
S	C	–
C	–	S
–	S	C

ANITA: *The deadlock is resolved! Search beats Ching twice, but Ching beats Search only once. By removing Min from the equation, we can suddenly see that Search is a better choice than Ching.*

NIGEL: *Yes...or perhaps it's the other way round: when we still had Min, his presence enabled us to see that Search and Ching are equally good! How can you learn more about two people by removing information, I mean, by removing a third person from consideration?*

STEPHEN: *True...Perhaps we have been barking up the wrong tree. Let's make a fresh start. I believe we do need to look at all three candidates: for all we know Min might decide not to withdraw after all. We should just abandon this silly notion that all three criteria are equally important.*

ANITA: *Good point. That's just Human Resources breathing down our necks. Surely all three candidates can do admin and teaching! We shouldn't use these things as selection criteria. We need someone who can teach, but we don't need the best teacher in the world. The same holds for admin. It's research that should make the difference. (The others agree.) On the other hand...what does count is someone's ability to work in teams. We need good team players here at Poppleton, don't we?*

STEPHEN: *We do—in fact, it's one of the criteria that we're meant to look at—but teamwork should count for less than research.*

NIGEL: *You know what? If one person is undeniably better at research than another, then that makes him a better choice, full stop. But if we're not really sure which of two people is the better researcher, then we give preference to whichever of them is the better team worker.*

STEPHEN: *Brilliant. Teamwork as a tie-breaker! Let's see how this pans out for our three friends. (Draws a new diagram.) I like your point, Nigel: if you twist my arm, I'm not so sure that Search really is better at research than Ching: the list of publications in his CV is longer, but only by the slimmest of margins. And similarly, I'm unsure that Ching really beats Min in terms of research. The only thing I am sure about is that Search beats Min. The difference in their track records is substantial.*

	RESEARCH	TEAMWORK
	S	M
	~	>
S > M	C	C
	~	>
	M	S

ANITA: *Hold on. It's the same old mess again! We prefer Search over Min, as you just said. But look what happens when teamwork is taken into account: in terms of research, we're unsure whether to prefer Ching or Min, so the tie between them is resolved by looking at their qualities as team workers. As it happens, this means that Min has the edge over Ching. (All nod.) Similarly, we're unsure comparing Search's research to that of Ching, so once again we have to break the tie by looking at teamwork, which tells us to prefer Ching over Search. So ... Min beats Ching and Ching beats Search, in other words Min beats Search. But we started off saying that it's the other way around! ... Each of the two is better than the other!*

NIGEL: *Oh dear...*

STEPHEN: *Folks, I'm getting tired. I propose that we all take a rest and reconvene tomorrow. (Sighs.) I sometimes feel I don't understand anything about people...*

ANITA: *Nothing to do with people. The same could happen when you compare cars.*

NIGEL: *Or football teams... When things differ along several dimensions, comparisons between them are tricky.*

Scientific discovery and word meaning

Looking back at our discussion of measurement, the role of scientific discovery is of particular importance. Our discussion of the concept of obesity, in particular, has shown that science has exerted a decisive influence on its definition. The philosopher Hilary Putnam has argued that the same is true for many ordinary words, such as 'water', for instance.[7] To get the idea, let us indulge in the kind of thought experiment that Putnam was such a clever inventor of.

Suppose you went to Mars and found a fluid there that looked and tasted like water. You might be tempted to take a swig. Suppose you are lucky and it relieves your thirst. Enthusiastic for more than one reason—water suggests the possibility of life!—you fill your bottle with the Martian fluid and bring it back to Earth. To your surprise, however, scientists inform you that your bottle does not contain the H_2O that we know and love, but some other type of molecule XYZ, unknown before your discovery. Will you say, 'We've just learned that water can consist of XYZ'? Putnam suggests that another response is more likely, namely to say that you've discovered *another* substance that shares many of its properties with water. In other words: even for such a common word as 'water', you ultimately rely on scientists to tell you what it means. You observe a substance, then ask scientists to find out what it is. In the case of water, scientists' verdict was to say that water is H_2O. From then on that's what the word means, perhaps until another definition gains the ascendency. Anything else may be fluid, transparent, and thirst-quenching... but water it is not.

What we have seen in this chapter is that the meaning of vague concepts is sometimes settled in the same way as that of Putnam's water, with an appeal to science. All the same, the point at the BMI scale where we call a person 'obese' is partly arbitrary. We have also seen that even where we attempt to make things precise, some residual vagueness tends to remain. We have discussed at some length why this is true for linear measurements, and the same is true for obesity because the BMI is ultimately based on the measurement of length. We shall often neglect such residual vagueness and pretend, for example, that 'taller than 160 cm' is a crisp concept even though there could be borderline cases. We shall return to this issue in Chapter 9. But first, let us turn our attention to some situations where vagueness and gradability play a more covert role.

Things to remember

♦ Despite centuries of efforts to give the *metre*, and all the measurement concepts derived from it, a precise definition, the metre is still a (very slightly) vague concept. The concept 'measures at least 1 metre', for example, admits boundary cases. This observation casts doubt on the standard division between vague and crisp concepts, suggesting a gradable notion (involving *degrees* of vagueness) instead.

♦ Scientific investigation can sometimes shed light on the question what a word 'should' mean. Methods for assessing obesity, for example, are informed by links with ill health and mortality. In cases such as the measurement of intelligence, where such objective anchors are largely absent, this method becomes problematic, and measurement is at risk of arbitrariness.

♦ Gradable concepts such as obesity and poverty are often accompanied by thresholds whose values are largely arbitrary and subject to change, causing them to become essentially crisp instead of gradable.

♦ *Absolute* measurements are often only a part of the story: to assess whether a person should be classified as poor, for example, the situation of other people in his or her community must also be taken into account. *Contextual* effects of this kind will play an important role in later chapters.

4

Identity and Gradual Change

We like the world neat and tidy. It consists of 'things', and these things have properties. Admittedly, some of these properties may be vague and fluctuating, but at least the thing itself is stable, isn't it? Surely, it is what it is and not something else.

We shall see that this rosy picture is not always easy to uphold. Let us dive in at the deep end, and look at a court case that hinges on the identity of a vintage racing car.[1]

Identity: The case of Old Number One

In July 1990, the High Court of Justice in London heard a case concerning the sale of an old racing car, known as Bentley's Old Number One. This British-made car was a vintage specimen which, in its time, had won famous races. In its most notable victory, in 1929, this very car had beaten off the American and German competition with such ease—slowing down deliberately as it approached the finish line, just to rub in its superiority—that it irritated the competition considerably. The court case of 1990 was noticed by the philosopher Graeme Forbes, who realized that it formed a nice illustration for some problems that spring

Cecil Kimber in 1925 at the wheel of the
first real MG Sports Car - The "Old Number One"
A Heritage Motor Centre, Gaydon image for editorial purposes only

BMR-0703-1182

FIG. 4 Old Number One

from the fact that the very identity of a thing is affected by vagueness. In a nutshell, here's why.

The buyer agreed to pay £10 million, and a car changed hands—but was it the Old Number One? The buyer thought it wasn't, and wanted to withdraw from the deal when he found out that the car he bought was considerably different from the thing that had won races. One of the experts called for the seller conceded that

none of the 1929 Speed 6 survives with the exception of fittings which it is impossible to date. Of the 1930 Speed 6 he believes that only the following exist on the car as it is now, namely pedal shaft, gear box casing and steering column. Of the 1932 car, the 4 litre chassis and 8 litre engine form in which it was involved in the fatal accident, he believes that the following exist: the chassis frame, suspension (i.e. springs, hangers, shackles and mountings), front axle beam, back axle

55

banjo, rear brakes, compensating shaft, front shock absorbers and mountings, the 8 litre engine, some instruments and detailed fittings.

To put it simply: so much has changed, that almost no part of the car has survived. Yet, the seller claimed that this is the authentic Old Number One considering, among other things, that 'racing cars are habitually changed during their careers, to incorporate improvements and modifications, and because of hard use and accidents'. The judge, weighing the evidence, reasoned as follows. First, he found that, after considerable rebuilding in 1932, 'it continued to be known as and was properly called Old Number One (...). There has been no break in its historic continuity from the time when it first emerged from the racing shop in 1929 until today.' The judge then clearly wrestles with the issues, arguing, for example, that

> The car is not and cannot be considered to be, or be known properly as the 'original' car which won either the 1929 or 1930 Le Mans. It would have to be composed of the same parts with which it left the racing shop or replaced by identical parts over the period of its existence, or the form for which it was prepared for the start of either race, or the form it was when it won. Degrees of originality, such as 'nearly original', 'almost original' or 'completely original', have no meaning in the context of this car. It could properly only justify the description of 'original' if it had remained in its 1929 Le Mans or Double Twelve form, even though such things as tires, radiator, fuel tank had to be replaced (more than once) due to the ravages of time or use.

In the same vein, he went on to argue that the car cannot be described as 'genuine' Old Number One. At this point the seller (who was the plaintiff in this court case) must surely have thought their case was lost. The judge, however, went on to argue that it was equally incorrect to describe the car as a mere 'resurrection' or 'reconstruction'. In the end, he came down squarely on the side of the sellers, deciding that the car could properly be described as 'authentic'. Talking about the car, he argued that

> At any one stage in its evolution it had indubitably retained its characteristics. Any new parts were assimilated into the whole at such a rate

and over such a period of time that they never caused the car to lose its identity which included the fact that it won the Le Mans race in two successive years. It had an unbroken period of four seasons in top-class racing. There is no other Bentley either extinct or extant which could legitimately lay claim to the title of Old Number One or its reputation. It was this history and reputation, as well as its metal, which was for sale on 7th April 1990.

Usually we do not worry about the identity of things. But the story of Old Number One, where a lot of money came to hinge on the way in which identity is understood, shows how difficult it is to say, of a changing object, at what point it becomes a different object. Few of us would have done better than the poor judge in this court case, yet his verdict becomes a muddle, in which such concepts as 'authentic', 'genuine', and 'original' all fall over each other to create an incomprehensible word soup.

Rather than rushing to conclusions, let us take a closer look at what happens when it's unclear whether two things are equal or not.

Multiplying objects

Jesus is credited with the multiplication of bread and fish. With the benefit of modern technology, let's see if we can go one better by producing multiple objects, each of which is *identical* to the other. Since we're talking about cars, let us multiply cars.

On our left we have the original, genuine, and unaltered Old Number One, which we met in the previous section. As part of a clever money-making scheme, we proceed as follows. We start by replacing one wheel, making sure to put the original wheel in a place well to the right of Old Number One. If we follow the judge's reasoning then this minor change certainly won't change the identity of Old Number One: it's still the same car. The next day we replace the second wheel. Since replacing a wheel doesn't change the identity of a car, we're still looking at Old Number

One. As before, we put the second wheel aside to the right, at just the right distance from the first wheel. The procedure is repeated for each wheel, and then for each of the other parts of the car. Whenever we replace a part, we make sure not only to put the old part aside, but to reunite it with the previously removed old parts in just the same way in which they were joined before, so that collectively they start to form a car. When we're done, then presumably the car on our left is still Old Number One. But, without a shred of a doubt, the car on the right is Old Number One as well. This one, after all, contains all the original parts, joined together into a functioning whole in the same way as before. We have produced two identical cars, each of which is worth millions, because each one is perfectly original.

Cars are made of rather dissimilar parts, some of which are rather large. Perhaps you feel that, once the chassis of a car is replaced, you're no longer looking at the same car. If that's how you—unlike the judge in the court case—feel then you might be swayed by a slightly altered example, which involves rowing boats rather than cars. Rowing boats consist of little more than a large number of planks, which are routinely replaced when they get mouldy. Yet, they tend to keep their name (which is often written on its outside) and are always thought of as the same boat. The thought experiment of the previous paragraph is easily repeated with rowing boats.

We have seen how an object can be changed step by step, using steps of such little consequence that none of them can be plausibly said to change the identity of the object. (After replacing one wheel, you're not claiming to be driving in another car.) But a series of small changes can collectively amount to a huge change, to such a point that the conclusion seems unavoidable that it's no longer the same object. We are reminded of the sorites paradox, which also appeared to hinge on an accumulation of small changes. What exactly is the connection?

The connection, I think, can be seen most easily by constructing a variant of the story of Old Number One that hinges on changes in just one dimension. Suppose somewhere in a museum, there lies a beautiful

Persian carpet. Centuries ago, it started out as the largest carpet ever woven, a feature that contributed to its fame and price. But the carpet has shrunk. As a result of some design flaw, little bits have fallen off, making it smaller every year. And now, in 2008, the carpet is about the size of a prayer rug. After three hundred years, the carpet is still beautiful but other carpets made at the same time, in other museums, are now much larger: they started out smaller, but didn't have the design flaw. The familiar sorites argument would try to prove that the carpet is still large. The analogous argument focusing on identity would try to prove that it's still the same carpet. (Needless to say, the thought experiment can be altered by shrinking the carpet even further or by involving other properties, such as its colour.) The connection is possible because, ultimately, an object is defined by its properties.

DIALOGUE INTERMEZZO
ON OLD NUMBER ONE

After the verdict in the court case on Old Number One, the protagonists might have carried on a conversation of the following kind. Unfortunately, there is little love lost between the gentlemen.

JUDGE: There's no doubt about it: the car you bought is the same car that won these famous races, eighty years ago. You'll have to pay up!

PLAINTIFF: The same car? Forget it. I am not going to pay. Now that its story is out in the open, there's not a living soul who would pay £10 million for this heap of rust.

JUDGE: Very sorry to hear that. If people are irrational then that's their business, not mine.

PLAINTIFF: Irrational? Your ruling is irrational.

JUDGE: Why?

PLAINTIFF: The core of your argument is continuity: the fact that the two cars are linked via a sequence of small repairs. But that's a ridiculous argument. Assume for a moment that we weren't talking about cars but books.

JUDGE: And your point would be...

PLAINTIFF: Let me finish. Suppose I have a book here. I now change it, replacing one word at a time. Deleting a word is also allowed of course. After one replacement, it's still the same book, right? After two steps it's the same book, and so on. Your argument doesn't take the number of steps into account, or the total amount of change, just the size of the steps.

JUDGE: Yes...

PLAINTIFF: But in that way, I can start out with the Bible and end up with The Pickwick Papers. According to you, that makes them the same book! Give me time for more replacements, and both will also be equal to Justine, Marquis de Sade's main claim to fame.

JUDGE: Come on, good man, we both know that these books are rather different from each other.

PLAINTIFF: I do, but you don't. It's you who claim that my car is the same as Old Number One, even though there's hardly a part that the two have in common.

JUDGE: Listen, you strike me as a reasonable man. You must have bought the car for your enjoyment, not just to sell it off. Why do you enjoy it less, just because you know more about its history? You can drive it the same way as before, polish it the same way as before, show it off the same way as before. Why the fuss?

PLAINTIFF: Because I want that car, that's why!

JUDGE: Come on, be reasonable. Did you really believe that no tyre had ever been changed in eighty years? If you don't repair a car every now and then, the darn thing stops working.

PLAINTIFF: Every now and then? Yes, of course. One repair fine, ten maybe, but there comes a point when you can no longer say that it's the same car. It's a matter of degrees!

JUDGE: But to let this depend on the number of repairs would be arbitrary. Where to draw the line?

PLAINTIFF: *Oh dear...You're not falling for that old chestnut, are you? You argue that any line would be arbitrary, and you conclude that the line can't be drawn anywhere. That's the sorites fallacy!*

JUDGE: *No philosophy please, this is a serious court of law! My point is that what counts is the idea of the car: the way in which all its parts form an organic whole. As long as the parts fit together in the same way, then it's the same car.*

PLAINTIFF: *You must be deluded. My grandson has toy models of cars, which work in much the same way as the originals. But even boys understand that a model is not the real thing. I'm giving up. I'm going to buy myself a famous painting, to console myself.*

JUDGE (inaudibly): *I hope you'll buy a counterfeit!*

What makes a book?

Few things change so dramatically during their lifetime as a book. Long ago I created a file on my computer and called it 'vague-book'. Since then, I have added many more words to it and rewritten every part several times. Now the book lies in front of me. Just like Old Number One, there is little that connects the end result with the starting point. Even the title has gone through several versions. In one sense it is still the same book, in another sense it is the last in a huge sequence of different books. If someone asked me how many versions the book had gone through, it would have been impossible to answer.

Translation introduces yet another dimension of variation. Suppose I tell you about the novel by Robert Musil, from which I quoted at the start of this book.[2] This novel is unusually complex. It must have been difficult to publish, because it was unfinished and unpolished when Musil died, and a nightmare to translate. Suppose one said the following about it:

[Musil's novel]₁ is justly famous. [The English translation]₂ was long overdue. There are places where [it]₂'s not very faithful to [the original]₃ perhaps, but [it]₂'s more easily digestible as a result.

In one sense, all of this is about the same book. But, as the numbers indicate, this is not the only way of thinking about the situation. What is *overdue*, for example, is the English translation (marked as 2), not the German-language version of the book (marked as 3). What is *famous* is something else again (marked as 1), namely the book regardless of the language. It is easy to imagine a series of translations, in different languages perhaps, each of which copies the errors in the previous translation and adds a few for good measure. The last translation in the sequence may be so badly botched that the meaning of the original book is changed beyond recognition. In a case like this, it would be difficult to say how many different books the series contains. Taking everything into account, the idea that a book is a well-defined thing is starting to look like a dangerous simplification.

Our earlier example, involving a famous racing car, was in some ways even more interesting because it involved something other than the things of which it is composed. What counted for the owner of the car was its history, so to speak. He may have cared deeply, for example, whether the 1990 car was the very thing in which a particular famous person had sat, which others had admired and stroked lovingly, and so on. The judge was right to invoke authenticity, a word more commonly associated with art, where emotional factors of this nature are crucial. This is nowhere more true than in Chinese art: if the painting in your attic is proved to have been owned by the emperor Qianlong, its money value surges to astronomical heights.

What makes a person?

But why talk about artefacts, if we ourselves are changing all the time? The cells of our bodies renew themselves through the years. Much is still

62

unknown, and the story is different for different parts of our bodies—it appears that the eye and parts of the brain do not renew themselves, for example—but recent estimates have it that the rate of change is so fast that the average age of all the cells in an adult person's body is only around ten years. Just like the case of Old Number One, each change triggers the question whether we are still the same person. But perhaps living things like ourselves are special, because we have a mind. One reason why I am still the same person as ten years ago, presumably, is that I remember that previous person. Memory enables me to carry out plans that I forged long ago. Memory helps to give us a kind of continuity that we would not have without it.

But if our mind is so vital, what if we were to lose it? If memory is so important, what if Alzheimer's disease strikes? The answers to these questions can be informed by research, in much the same way as medical research has given us tentative answers about the measurement of obesity. Answers of this nature, however, are almost unavoidably tentative and dependent on one's view of life. I doubt that any definite answers will ever be forthcoming, and in this sense the vagueness surrounding the notion of personal identity is probably here to stay.

Needless to say, the same issues are at play in connection with the start—and the end—of life. For suppose we were somehow able to defend the usual position that the *me* who writes this is the same as the *me* who was entered into the birth registry about half a century ago. What does this mean for the *me* just before birth, let alone 24 weeks into my mother's pregnancy? Here, of course, we are getting into hotly debated territory, with quite a few people arguing that this foetus is essentially identical to the later me, while a majority in the British House of Commons believe otherwise, presumably because this early foetus is too different from me to count as me.

To top it all, the boundaries of an organism are unclear even at any given moment. Examples in the philosophical literature include a cat that is about to lose a hair. Is the hair part of the cat? And what of the bacteria in its gut: are they part of the cat or something separate from it, even though it needs them to survive? Similar questions can be asked of people, whether we like it or not.

What is a language?

There are said to be six to seven thousand languages in the world. People like to quote this figure, but it is thoroughly unclear what it is based on. Languages, like everything else, can be counted only if we have an identity criterion. Consider our garden, which is regrettably full of moss. How many moss plants are there in it? Without an identity criterion, I couldn't tell you where one moss plant ends and the next one starts, so I couldn't give you the answer. In the same way, it is far from clear how many languages there are in a given part of the world until we know how to tell languages apart.

What would be a reasonable criterion for finding the boundaries of a language? Are British English and American English just two dialects of one language, or should they be counted as different languages? Usually people don't care about such questions, but things can become different when an expensive new dictionary project is looking for funding... or if a conflict is in the offing. The point is sadly illustrated by the insistence by some in the Balkans that what was once known as Serbo-Croatian is, on closer inspection, a set of two different languages, Serbian and Croatian. Nothing sharpens the linguistic mind like a military conflict, apparently.

Some argue cynically that a language is just 'a dialect with an army and a navy'. Others argue that two people speak the same language if they can understand each other. There is something pleasantly principled about this move, because it places communication, as the purpose of language, in the driving seat. It does, however, lead to some unusual conclusions. For suppose we lined up all speakers in a huge queue, where each person speaks very similarly to the next person in the line.

Then, exactly like the *species* discussed in Chapter 2, speakers 1 to 10,000,000 might all understand each other, but so do speakers 2 to 10,000,001, speakers 3 to 10,000,002, and so on. It follows that there exist millions and millions of different languages, each of which overlaps with many other languages in terms of its speakers. Intellectually defensible though this position might be, it is not how we usually think about languages.

We have seen a number of examples where small differences make it difficult to judge whether two things should count as the same or not. In many cases, the differences involved arise through time. Given enough time, the thing might become unrecognizable. Or maybe it won't, because the thing is protected against change.

Digression: Protection against change

Computer scientists know all about change. They call it data corruption. Computers store information in the smallest units imaginable, known as bits. A bit can be on or off, telling you that something is true or false. When written on paper, 1 traditionally stands for yes (true), and 0 for no (false). Internally, at the level of hardware, the difference between the digits 0 and 1 is encoded as a difference in voltage: at this stage, the difference is analog, not digital; the thresholds for what counts as high and low voltage are chosen just far enough apart that minor variations in voltage will not cause bits to change into their opposite. Essentially, the quantization step from voltages to digits is what gives digital technology its edge over analog techniques, because digits are robust and easy to perform calculations on. Yet errors can and do happen, for example because a transistor is getting too hot, or simply because of wear. This is particularly dangerous when the information that is stored or transmitted does not contain much redundancy. Let me explain.

Suppose you want to transmit someone's phone number. Computers represent numbers using strings that consist of the above-mentioned bits, instead of the ten digits of the more familiar decimal notation. This *binary* representation uses the bitstring 0 to represent the number 0, 1 for the number 1, 10 for 2, 11 for 3, 100 for 4, and so on. Each string can be preceded by as many zeros as you like. Just like English words, phone numbers obey certain regularities. Yet, there is room for errors: if the last digit of my phone number gets corrupted from a 9 to a 7, you'll get to talk to one of my colleagues; if the last but one is corrupted you may get someone in a different department. In such cases, the problem is not just that an error was made: there is no way for the computer to tell! Luckily, computer scientists have invented schemes for reducing the likelihood of errors. Let us look at two such schemes.

Suppose you want to send a bitstring 0110, minimizing the likelihood of problems. One option is repetition: you send the same bitstring a number of times. In this way, you make sure that the original string is recoverable as long as the error does not occur too often. If, for example, every string is sent three times and the arriving signal is 0110 0110 0111, then majority rule allows you to conclude that the third part of the triple must be incorrect (or else there must have been two errors, of exactly the same kind, which is less likely). But, approaches based on repetition are expensive, because all information must be sent several times.

A more economical approach is to use what is known as a *parity bit*: you add up the bits in the string, and add one extra bit to say whether the sum is even (0) or odd (1). In our example, the bits in the original string 0110 add up to 2, which is even, so you add the bit 0, resulting in 01100. If the same transmission error occurs as before, the arriving string will be 01110. Now the computer performs a simple check, to see whether the first four bits sum to an even number (as the parity bit suggests), which is not the case. Now the computer can tell that something has gone wrong, and ask for the information to be re-sent. This approach is less costly than the previous one. It can detect only *odd* numbers of errors, but no error correction scheme can be foolproof.

These are just the simplest elements of a vast array of computational techniques designed for the automatic detection—and often the automatic

66

correction as well—of transmission errors. Together, they ensure that the messages in your email box are likely to reflect what their senders have written.

Computer scientists are not unique in their concern with data corruption. Look at the letters of our alphabet, for example. They may seem like mere scribbles on a piece of paper, but there is more to them. If you copy something I wrote, your letters will be shaped differently from mine, but they're part of a sophisticated convention. Once you know the conventions, you know that the precise length of the vertical bars is irrelevant, for example: all you need to know is whether the bar is short (as when you write an *a*) or long (as when you write a *d*). The effect of all these conventions is that we can often read each other's handwriting. Moreover, texts do not get corrupted over time so easily.

Suppose you find an old manuscript, which is written in English but rather the worse for wear. You may have difficulty deciphering some of the letters, but each one can represent only one of twenty-six options. These twenty-six letters are just about as different from each other as possible without making them too complex to write. The same is true for higher levels of information, for example at the level of words. It has been shown, for example, that an English text in which only the first and the last letter of each word is legible is often still intelligible to a human reader. Broadly the same principles apply to the intelligibility of speech: there is huge variation in pronunciation, yet we do understand each other, much of the time.

Richard Dawkins has written insightfully about DNA as a self-normalizing code in biology. Being based on an alphabet of four different proteins, DNA is self-normalizing in much the same way as words, preventing DNA from corrupting when passed on from parents to children. He has even applied the same principles to the transmission of ideas and designs (*memes*) between people. To see what he has in mind, consider the example of a car again. Suppose you found the only remaining specimen of a forgotten type of car, from the 1900s maybe. There may be dents in the exterior, and rust in some places, and one or two parts may be missing from the motor. Would you have been able to reconstruct the original car, faithful to the original in all essentials (mechanical structure, outward appearance, maximum speed, etc.)? As long as the damage to the car is not too bad, the answer will tend to be yes. The rust can be removed, dents undone, and a few nuts and bolts replaced—only incidental properties such as colour remain uncertain. This reverse-engineering is possible because cars have a predictable structure: try to connect a nut with the wrong bolt and you fail, so you'll keep trying until you've found a match.

The principle in all these cases is that information tends to be encoded redundantly. The result is that, if a small error occurs, the recipient will be able to detect and correct it, thereby preventing a 'Chinese whispers' sequence in which the message is corrupted further and further until it is beyond repair.

So what?

After this small detour, let us return to the problematic cases with which this chapter started. These examples suggest that the world around us is all fluid, like the proverbial river into which no one can step twice. Perhaps we need to give up thinking about the world in terms of objects that have properties.[3] A *mountain* is a good example. Even with a prominent mountain such as Mont Blanc, drawing boundaries around it would be highly artificial. Much better to abandon talk of mountains, speaking of differences in altitude instead. The same is true for wind and rain. There are differences in barometric pressure, and they cause movements of air. That's all. There is no wind. The proposal is to abandon all loose talk. Far from saying that my former colleague Milo speaks Serbo-Croat, we will say that the cells currently making up the person who used to be my colleague speaks in a way that makes him understandable to this particular set of people—I mean by these other conglomerates of cells.

68

For most purposes, however, this would be a bad move. Different models of the world serve different purposes. It is true that a geologist might, when she is performing a particular kind of calculation, do away with mountains, replacing them by differences in altitude. But most people benefit from the simplicity of ordinary language, which is cast in terms of mountains, wind, people, and the rest of it. It would be foolish to sacrifice utility for precision.

In the remainder of this book, we shall keep things simple by pretending that the world is made up of *objects*. This simplification will allow us to get the vagueness of *properties* (also called qualities, predicates, or features) more sharply into focus. Before turning to the linguistics and the logic of vague expressions, let us briefly examine an area where vagueness may be least expected: numbers.

Things to remember

◆ Most of the things around us have boundaries that are only vaguely defined. This is true for cars and boats, for example (because their parts tend to be replaced), but also for ourselves (because we age). Cases arise where it is unclear whether we are still dealing with the same thing.

◆ These problems are most evident in connection with gradual change, particularly when the 'thing' in question is very long-lived (e.g. a language), or when the rate of change is high (e.g. a book that is being written).

◆ Parity bits and related schemes in computer science can be seen as fallible safeguards against gradual change.

◆ To regard vaguely defined events as if they were crisply defined 'things' is perhaps best seen as a useful fiction.

5

Vagueness in Numbers and Maths

Numbers, pure and precise in the abstract, lose precision in the real world. It is as if they are two different substances. In maths they seem hard, pristine and bright, neatly defined around the edges. In life, we do better to think of something murkier and softer. (...) Too often counting becomes an exercise in suppressing life's imprecision.

BLASTLAND AND DILNOT 2008

Numbers, in my case, are an acquired taste. As a schoolboy, I knew that numbers have a role to play in life. But I felt that, most of the time, what counts is something else. To determine whether the distance from the railway station to our house is best travelled on foot, you do not need to know the exact number of metres: to know that your little sister has walked the same distance last evening is sufficient. Numbers do not come into this. If you asked me to put a figure on the distance, my approximation could be off by as much as 30 per cent or so: for practical purposes such a margin is often acceptable. Even questions about numbers can often be answered without laborious calculation. When you're asked which of two shopping lists adds up to the greater total expenditure, you may not need to add: it would be enough to observe that each number in one list can be paired with one in the other, where the number in the second list is at least as large, while at least one of them is larger. In examples of this kind, it's not the numbers themselves that count, but comparisons between them.

I have now grown up and learned a few things, and my interest in numbers has grown. Our discussion of poverty and obesity, for example, would have been impossible without numbers and measurement. Indeed, even approximations can involve a considerable amount of numerical calculation.[1] In the present chapter, we shall examine numbers more closely, asking whether they are always as exact as they seem. In doing so, we shall also be looking at vagueness in mathematics more generally, which is the natural habitat of numbers, after all.

Vagueness in mathematics

Years ago, as a student at the University of Amsterdam, I took a course on Gödel's famous incompleteness theorem, a fiendishly difficult subject. The room I entered at the start of the first lecture, along with four or five other brave students, was a perfectly normal lecture theatre, with a blackboard of about normal size. Before long, the lecturer entered: Professor Martin Löb, an elderly gentleman who, in his time, had proved important new theorems on the subject of the course but who, like quite a few logicians, was famously inarticulate. He stood there for a while, looking rather helpless. Then he began to speak, softly and apologetically: 'The eh, hm, blackboard is too small. The formulas in this area tend to be, hm, eh, rather long', after which he disappeared. After five minutes he reappeared, asking us to follow him to another hall, in the bowels of the building, where few of us had ever ventured. The hall was huge. What was worse, its vast front wall was covered over its entire surface by enormous movable blackboards! I hadn't felt so intimidated in my life. Evidently, vague descriptions can matter to a mathematician: descriptions of the size of a formula for example. One wonders where else they can be relevant.

71

It is well known that it took the world a long time to figure out that zero is a number worth having and to work out its properties. The same is true at the other end of the scale, involving large numbers. A well-documented case is that of expressions such as *murios* or *murioi*, in ancient Greek. These Greek words are variously translated as 'ten thousand', 'many', 'countless', or 'infinite'. It simply isn't always clear—to us, and very possibly to the ancient authors themselves—what the expressions involved mean. Presumably, the ancient Greeks had little use for seriously big numbers, since the largest sets worth measuring, in those days before mega- and terabytes, must have been armies, and armies of a happily limited size at that. Although languages such as modern English have long had words for larger quantities, it is interesting that the English word 'infinite' appears to have functioned in the same way as *murioi*: a few centuries ago, the word 'infinite' had only a vague meaning, in which it meant something like 'more than can be counted in the time available'. A number of people have contributed to the modern conception of infinity, but it was not until around 1875 that the mathematician Georg Cantor came up with a general and rigorous analysis.[2] This analysis gave the word 'infinite' an exact interpretation, essentially by deleting the words 'in the time available' from the phrase above.

The upshot of Cantor's analysis—which was initially controversial but stands essentially unchallenged today—was to declare all sets (and only those sets) infinitely large whose members are too numerous to *ever* be enumerated exhaustively: any number in the set $\{0, 1, 2, \ldots\}$ would underestimate the size of such a set. For example, the set of all positive integers smaller than a *trillion* is not infinite because if time is not an issue its elements can be enumerated one by one. What we would find, of course, is that its size is a trillion minus 1, which is not infinitely large. For the same reason, the set of all the particles in the universe is almost certainly finite. It takes a while to count them, but it can be done in principle. The set of all even numbers, however, is infinite according to Cantor's definition, and so is the set of all numbers greater than 100,000, for example. After Cantor, it became possible to use infinitely large numbers in precise calculations. The details are of no concern to us here.

As a result of clean-up operations of this kind, vagueness has almost been eradicated from the subject matter of mathematics. Yet, vagueness continues to play a huge role in the presentation and explanation of mathematical results. This is perhaps most evident in the way mathematics texts tend to be organized, around main *theorems*, *lemmas*, and *corollaries*. All these are things that are proven, and the difference between them is typically defined along the following lines.

> Formally, a theorem is a statement that can be shown to be true. (...)
> A *less important* theorem that is helpful in the proof of other results is called a lemma (...). Complicated proofs are usually easier to understand when they are proved using a series of lemmas, where each lemma is proved individually. A corollary is a theorem that can be established *directly* from a theorem that has been proved.   (Rosen 2007: 75; italics are mine)

These definitions are vague, because it is not specified exactly how 'important' a proposition needs to be before it can be called a theorem, and how 'directly' a corollary needs to follow from a previously established theorem. So, by using lemmas and corollaries, vagueness is built implicitly into the very structure of a typical mathematical text. Sometimes vagueness plays a more explicit role. To see how, here is a classic article by the mathematician Thoralf Skolem, which contains vague expressions at all its junctures. I summarize it, taking care to omit all mathematical content:

> In volume 76 of *Mathematische Annalen* Loewenheim proved an *interesting and very remarkable* theorem on (...). Loewenheim proves his theorem by means of (...). But this procedure is *somewhat involved*. In what follows I want to give a simpler proof (...) It is *immediately clear* that (...) But it is now *easy to see* how (...). *Strictly speaking* (...) We can then *easily* convince ourselves that (...). It would now be *extremely easy* to prove this theorem in Loewenheim's way (...). However, I would rather carry out the proof in another way, one in which we (...) proceed more

directly according to the *customary* methods of mathematical logic. (...)
Now let X^o be (...) Then, as is *well known*, X^o is (...) But it is *clear*, further,
that (...) must hold. (...) By *well-known* set-theoretic theorems it then
follows that (...) The same method of proof can, of course, also be
employed if the structure of (...) is still more complex. We could, for
example, prove a *very similar* theorem about (...).   (Skolem 1920)

Over and over, vague language is used to indicate the difficulty of a
procedure. The frequency of vague expressions in mathematical texts
suggests that vagueness plays an important role in people's thinking
about mathematics, even if the concepts involved are crisp. Presumably
readers use vague information to understand what the author has in
mind. When reading that a proof is 'simple', for example, it becomes
easier for us to reconstruct it, because we can disregard complex proofs.
This is a bit like a chess problem in the newspaper: by instructing us to
find a 'checkmate in three moves', for example, the newspaper helps us to
find the winning moves, because we do not have to consider long-winded
solutions. It is by dropping hints of this kind that the newspaper allows
us ordinary readers to find a solution to a problem that may well have
eluded a grandmaster when it occurred in a real game. The difference is
that chess moves are easier to count than the steps in a mathematical
proof, which is why, in mathematics, readers usually have to make do
with vague hints.

Perhaps then, vagueness plays a role only in the elucidation of mathematical constructs, not in the contructs themselves. Although this is
probably the standard view of the topic, it will be worth paying attention
to one interesting exception, which is associated with the term *strict
finitism* (also *ultrafinitism*). In our explanation, we shall broadly follow
the discussion in a famous article by the philosopher Michael Dummett.

Strict finitism (Essenin-Volpin 1970) is a late response to the problems
that plagued mathematics in the beginning of the twentieth century and
which have never entirely gone away. To reduce the risk of formulating
mathematical theories that are essentially faulty, finitism bans all infinite
objects from its calculations. If a set is infinite, many things become

difficult. You can never use a computer to check all its elements for some property one by one, for example. Ultrafinitism is even more radical, allowing only those mathematical constructions that can *actually* be carried out. It does not have many adherents at the moment, but it will be instructive to examine its implications.

Consider a number like 10^{100} (a 1 followed by a hundred zeros), for example. This number, which is also called 'googol' and is almost certainly greater than the number of particles in the universe, is easily too large to be enumerated in a lifetime: it cannot in practice be contructed by starting from zero and adding units (i.e. adding 1) again and again. But if exponentiation is allowed (i.e. raising something to the power of, say, 10) then one or two seconds suffice to construct the number. Thus, what numbers count as acceptable depends on the operations that the finitist allows himself to use. The point, however, is that, regardless of the set of operations, strict finitism implies that not all natural numbers can be acceptable. For there exist infinitely many natural numbers, so if these were all acceptable then one would be forced to accept the existence of at least one infinite set, which, to the strict finitist, is the road to disaster.

Confronted with this conclusion, the strict finitist can either reject all numbers out of hand—not a very tempting option of course—or draw the line at some arbitrary point: she might regard numbers up to some fixed number n as acceptable, but all higher ones as unacceptable. The finitist does not have to be picky, of course: she could draw the line at 10^{100}, but she could just as well choose $10^{100} + 1$: a value of 1 will not make any difference. In other words, in reasoning about finitism, the acceptability of a number is a vague concept! This fact could be exploited to reach a conclusion that the finitist cannot accept, because it gives us back ordinary mathematics:

o is acceptable.
If n is acceptable then $n + 1$ is acceptable.
Therefore every natural number is acceptable.

(The reasoning employed here is known as *mathematical induction*.) The situation is mirrored in computer science if we ask which natural numbers can be efficiently represented on a given computer.[3] The largest of these numbers is often called *maxint*. 'Maxint' is an important concept, because errors can result if a computer program forces the computation of a number larger than maxint, for example, by asking it to calculate $2 \times$ maxint. Now suppose a computer manufacturer claimed that their computers are able to represent numbers of any size. To prove their claim, they might observe that, given a computer whose maxint equals n, they can always build one whose maxint is $n + 1$ and conclude, along sorites lines, that *any* natural number can be manipulated by their computers. The plausibility of this claim could be argued to depend on how exactly it is read: computers using *any* maxint might in principle be built (given enough time and resources); but, at any given time, there is a largest maxint (i.e. a number n such that no computer has a maxint higher than n). Philosophers have a term for this idea: the set of numbers representable is *potentially*, but not actually, *infinite*.

Computers' limited space for representing numbers also means that fractions often have to be rounded. Rounding can be seen as a small error in the representation of a number. But in 'chaotic' systems, small errors can sometimes be magnified, as when the proverbial butterfly flaps its wings, causing hurricanes. Here is a miniature example of how this can happen in a computer program. Suppose the number 0.999 was rounded off to 1. In most situations, a rounding error of $1/999$ would be acceptable, but if the number in question is then multiplied with itself a large number of times (because this happens to be what your computations require), then a small error turns big. 1^{5000}, for example, is just 1, but 0.999^{5000} is 0.0067. The further we go, the larger the error gets, causing the number to approximate 0 ever more closely: 0.999^{10000} equals a mere 0.0000452, for example. Far from being a specific fact about computers, of course, the point is a general one which shows how important precision can sometimes be, and how hazardous a lack of it. To be told that your chance of winning the lottery equals the result of raising a number close to 1 to the power of 10,000 would not be very informative!

Talking about numbers

(...) the museum guard (...) told visitors that the dinosaur on exhibit was 9,000,006 years old (...). Upon questioning, the guard explained that he was told the dinosaur was 9,000,000 years old when he was hired, six years before.   (Paulos 1995)

It is tempting to think of ancient times, when numbers such as 'infinite' could lead a somewhat shady existence, as a primitive stage of language development. But perhaps this temptation must be resisted. For even though numbers can make information precise, in many cases they have to be taken with a pinch of salt. Let us see why, by looking briefly at the way scientists talk about numbers, before we direct our attention to ordinary language.

Scientists at work are not qualitatively different from anyone who is measuring up a room, since most of the factors mentioned above can also affect the scientist: unclarities as to what it is that is measured, errors in carrying out the measurement, and the lack of precision inherent in even a correctly executed measurement. The difference between a scientist and a layperson is largely a matter of degree: where you write 476 cm, a scientist might write 476.3724 cm (or something equivalent to it) if she wants to be really precise. Although this makes the scientist more precise than you, it does not make her *completely* precise: both measurements have a margin of error. The only thing setting apart the scientist from the handyman is the fact that scientists think systematically about their margin of error. If a scientist writes '476.3724 cm' then her claim is that the actual value should lie between 476.37235 and 476.37245.

We have talked about two kinds of 'scientist': the real one, in her laboratory, and the mock scientist who is measuring up a room to fit a carpet. Both cases are exceptional because they involve a person who is trying to be precise. In ordinary life, numbers are often treated rather more shabbily, as we shall see.

First of all, ordinary language uses some conventions of its own. For example, when stating an adult person's age, we tend to round all digits down: a person who was born 47 years and 250 days ago counts as 47, not 48 years old. This does not usually lead to unclarity, except when the person is very young, in which case we are sometimes more precise: when asked for the age of a newborn baby, we don't say that it is 0 years old (as the normal conventions demand), but rather that it is 6 weeks old, for example. Unclarities can arise when we talk about toddlers, who are somewhere between babies and adults: when we say that a toddler is 2 years old, do we mean to say that her age is between 2 and 3 (as with adults), or do we mean something more precise?

But the case of a person's age is probably exceptional, in that the conventions that we are using are fairly well defined. By comparison, suppose we are talking about a sum of money that was stolen from the bank yesterday, which was reported by newspapers as amounting to '27 million'. What is the margin of error? Are we talking about 27,000,000, or would 27.10^6 (using scientific notation) better reflect the (lack of) precision involved?

Somehow, a number like 27 million is special to us, because it is nicely 'round'. The linguist Manfred Krifka (then at the University of Austin at Texas, now at Berlin's Humboldt University), investigated the issues surrounding round numbers and came up with interesting results. For example, he observed that in many languages, multiples of ten (and less so, multiples of five) tend to be expressed more briefly than other numbers. 'Ten', 'twenty', and 'a hundred', for example, have fewer letters than 'nine', 'nineteen', and 'ninety-nine' respectively, and also than 'eleven', 'twenty-one', and 'a hundred and one'. To explain what goes on when we interpret numbers, he argued that there are two principles at work here, neatly summed up by the title of his paper: 'Be brief and be vague!' (2002). He contrasts two statements:

(a) The distance between Amsterdam and Vienna is one thousand kilometres.
(b) The distance between Amsterdam and Vienna is nine hundred and seventy-two kilometres.

Krifka observes that (a) tends to be interpreted vaguely but (b) much more precisely. For example, the response 'No, you're wrong, it's nine hundred and sixty-five kilometres' would be odd in response to (a) but not in response to (b). To explain this, he postulates his two principles: to be brief, and to be vague. These principles are not cast in stone: they are meant to tell us what is normal. The following pattern emerges if we list the different ways (i.e. a vague and a precise way) in which each of the two sentences can be interpreted, one of which is brief and the other lengthy:

		BREVITY	VAGUENESS
1.	a [brief] [vague]	normal	normal
2.	a [brief] [precise]	normal	abnormal
3.	b [lengthy] [vague]	abnormal	normal
4.	b [lengthy] [precise]	abnormal	abnormal

If one buys into Krifka's two principles[4] then it is obvious that (1) gives a better account of (a) than (2), since (1) says that the speaker has chosen a 'normal' expression (i.e. a brief one) to express a 'normal' content (i.e. a vague one). Where does this leave (b)? One might argue that, once again, it is best to assume that the speaker intends to express vague information since this is the normal thing to do. This, however, is not what we want, since (b) tends to be interpreted as precise rather than vague. To obtain the desired prediction, Krifka invokes a rule which says that normal goes with normal, but abnormal goes with abnormal. Applied to our example (b), this means that the abnormality of its form—it is abnormally lengthy!—implies abnormality of interpretation as well, hence the precise interpretation.

The vagueness of round numbers has interesting practical implications when a round figure in one language needs to be translated into another. Suppose, for example, the English user manual of a kitchen hob says 'This oven should not be installed within a yard of a refrigerator'. Suppose you need to translate this into Dutch, which does not use the yard. A literal rendering would translate 'a yard' into '0.9144 metre', but this would greatly exaggerate the precision of the original statement. A sensible translation would replace 'a yard' by the equally brief expression 'een meter' ('one metre'), trusting that the difference between the two is too small to matter in this particular situation. But what if the situation requires more precision, in an industrial setting perhaps? Or what if no near-synonym (like 'metre') is available? This happens when the original English text is cast in terms of miles, for example. Replacing 'a mile' by 'een kilometer' is possible, but since a mile is about 1.6 times as long as a kilometer, the difference tends to be too large to neglect. Newspapers have guidelines for dealing with tricky issues of this kind, but a silver bullet does not appear to exist.

In this section, we have been able to show only the tip of the numerical iceberg: actual talk about numbers involves interesting mixtures of numerals and ordinary language, as anyone who listens to weather reports knows. ('Temperatures will be in *the low twenties* today', or even 'Temperatures will *struggle to reach fifteen*'.) What counts, for us, is that while numbers are precisely defined mathematical objects, statements involving numbers are usually not. Speakers and hearers understand this, and they constantly exploit the subtle rules that govern the use of numbers. These rules are seldom addressed in dictionaries, grammar books, or language guides, probably because these issues are common to all languages, so that second-language learners do not need to be told.

Which computer program is fastest?

Computer scientists often compare programs, asking such questions as 'Which one is fastest?' To see what such questions amount to, let us step into that fulcrum of rational thought: the betting office.

Suppose you want to bet on the outcome of a football match. You find three betting offices. Losing the bet means losing your entire stake. But if you win, one office offers you five times your stake, the second 100 times, while the third one offers you five times the square of your stake. Which one has offered you the best deal?

The Royal Ripoff: $5n$
SmallBets Inc.: $100n$
BigSpenders: $5n^2$

Clearly, Ripoff's offer is inferior to the others, each of which gives you a higher return regardless how much money you put in (as long as n > 1). But how can the other two be compared? It depends how much money you're willing to stake. In the following table, n is the number of pounds you might be betting, and the numbers in the cells represent the number of pounds paid to you if you win in each case:

	$n=5$	$n=10$	$n=20$	$n=30$	$n=40$
Ripoff	25	50	100	150	200
SmallBets	500	1,000	2,000	3,000	4,000
BigSpenders	125	500	2,000	4,500	8,000

SmallBets is best for some bets, but BigSpenders is best for others. This suggests that comparing functions is meaningful only if you know what value they will be applied to (i.e. how much money you're going to stake). In other words, it depends. In many situations, however, such a non-committal position is not satisfactory. Computer science is a source of many such situations.

Computer scientists often need to choose between programs that perform the same task using different strategies. Different considerations come into play, such as the amount of memory that these programs require, or their elegance. (It's a bit like planning a trip: you may be interested in finding the fastest route or the most scenic one.) One important criterion for choosing between computer programs is the *time* they take. In many situations, one would like to choose the quickest program, but what does 'quick' mean? As with the betting office, one program may be quicker on some inputs while another program may be quicker on others. What one would like to know is which program is quickest when all possible inputs are considered.[5]

Let's be more concrete. Suppose the programs we are talking about perform the task of putting lists of names in alphabetical order. You do not know in advance how long the lists will be. Your programs might behave just like the three betting offices above, where n now represents the length of the list, and the numbers in the table the number of milliseconds your computer will run in the worst case (e.g. if your initial list is far removed from alphabetical order, so a lot of reordering work needs to be done). Many computer programs grow even more rapidly than the ones discussed so far: it is not unusual to see programs associated with a growth rate of 2^n, where n measures some feature of its input such as, for example, the length of a list.

Computer scientists' need to compare functions of this kind has given rise to a beautiful mathematical theory, the theory of *computational complexity*. A few of the most basic principles of this theory are worth sketching here informally. As so often happens when scientists design metrics, they remove vagueness while introducing a certain arbitrariness.

The first idea regards the way in which complexity is *measured*. The idea is to stipulate that one function is 'greater' than another if and only if there exists a point beyond which its values are greater than those of the other one. Consider the functions $100n$ and $5n^2$, which were discussed above. We can now say that the latter is greater than the former because there is a point, at $n = 20$, *beyond* which $5n^2$ is always greater than $100n$.

The motivation behind this stipulation is that small n (e.g. short lists of names) are usually no problem: it's large n that merit scrutiny. How large? Arbitrarily large!

Like every simplification, this one does not *always* give us what we want. What if one function starts producing larger numbers than another one only for n that are so ridiculously large that we shall never encounter them in practice? We have reached one of our slippery slopes again. Consider the names-sorting task: it is reasonable enough to regard a program that takes $5n^2$ time as slower than one that takes $100n$ time, because the programs will tend to be used on lists that contain many more than 20 names; the same is true for $1000n$ because, in many applications, lists will tend to contain more than 200 names. But what about $10^{100}n$? Only lists that contain more than 2.10^{99} names would be long enough to justify regarding the $5n^2$ program as slower than the $10^{100}n$ one, and lists of such staggering size are unlikely to occur in practice. The theory of computational complexity tends to disregard such trivialities and counts the former program as faster because *there exists* a number n beyond which $10^{100}n$ is smaller than $5n^2$. Never mind how large this number is. This principled stance keeps the theory neat and tidy, but this can sometimes be at the expense of practical utility.

The second idea is reminiscent of *rounding*, except that it is much more drastic. Simply put, the idea is to disregard all constants! Consider two programs whose run times are associated with the functions $5n$ and $100n$, as in SmallBets and the Royal Ripoff above. These two functions are equal except for the constants, 5 and 100. Computers are getting faster all the time: what takes today's computers 100 seconds will soon take only 5 seconds, thereby obliterating the difference between these two functions. (If history is a guide then this will take about a decade.) In this light, constants are peanuts. Functions that differ only in their constants are therefore treated as being of the *same size*.

Together, these two ideas have given rise to a wonderfully abstract way of classifying programs in a hierarchy of broad classes. The Ripoff and SmallBets functions, for example, belong to the unproblematic class of *time-linear* programs, BigSpenders represents a slightly more time-intensive class of *time-polynomial* programs, and a function like 2^n exemplifies the class of *time-exponential* programs, which are generally regarded as problematic, because they grow so rapidly. Programs' consumption of memory is quantified using the same broad brush strokes. In both cases, the pattern is essentially familiar to us from our discussion of obesity: one can talk vaguely about slow computer programs and fast ones, but a specially designed mathematical theory allows us to replace vague talk by crisp categories. As in the case of obesity, the price we pay is a certain arbitrariness in the categorization, which might occasionally make us a bit unhappy about the way in which a particular program is categorized.

In its full glory, the theory of computational complexity does not look only at specific computer programs, but also at the problems themselves that these programs are meant to tackle, trying to assess how quickly they could be solved by *any* program, or at the cost of how much computer memory. A discussion of these more abstract matters, however, would distract us from our main theme.

Statistical significance

The *New Penguin English Dictionary* lists several meanings for the word 'significance', one of which is aptly split in two:

1. having or likely to have influence or effect; important.
2. probably caused by something other than chance.

We shall be concerned with the second of the two senses, which is a cornerstone of the empirical sciences. It is known as *statistical* significance and was invented in the eighteenth century by Pierre-Simon Laplace and others. Essentially, statistical significance is a way of using numbers to measure the rashness of a generalization.[6] We shall see that, like its sister concept (1), statistical significance is gradable, even though it is often misrepresented as crisp. By way of an introduction, here are a few words about statistics in general.

Descriptive statistics is a modest but solid endeavour: essentially, it is the business of summarizing known data, for example by saying what its range (i.e. minimum and maximum value) is and its average. The annual report of a company is an exercise in descriptive statistics. As long as no errors are made—and this is a huge caveat—descriptive statistics is the domain of boring certainties.[7] Statistical significance is the cornerstone of *inferential statistics*. Inferential statistics involves taking some data and tentatively inferring (i.e. extrapolating) something about a larger chunk of the world. Suppose, some time during the Cold War perhaps, an American army general wanted to know whether soldiers in the USSR were taller, on average, than American ones. Suppose he was sitting on a lot of descriptive statistics concerning the heights of American soldiers, whose average height turned out to be, say, 175 cm. He knew all about his own army, but what he didn't have was data on the enemy. He had a dim suspicion that they might be taller. How might he test this hunch of his?

Given that even a general does not have unlimited resources, his natural course of action would be to compile a limited 'sample' of data. Suppose, for example, he could measure only *one* Russian, chosen randomly of course (i.e. not from the members of a basketball club). Suppose this person's height was 177 cm, 2 cm taller than the average American soldier. Before pondering the potentially devastating consequences of his finding (Russians are taller than Americans!), however, he realized that a sample of one may not reflect the truth about Russians in general. His finding may have come about by accident. To jump to conclusions about Russian soldiers in general—and to claim something about things you haven't measured is *always* a jump!—is the type of extrapolation step that gives inferential statistics its name. To make a step of this kind with confidence, one has to perform a statistical test. Such a test might have allowed the general to say whether the difference in favour of the Russians is *statistically significant*.

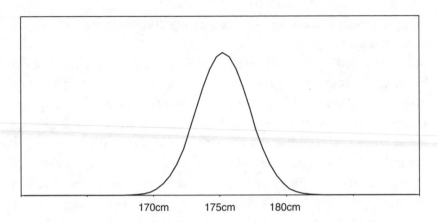

170cm   175cm   180cm

FIG. 5 Relative frequencies of heights among American soldiers (on fictitious data)

A statistical test has to take various things into account. In addition to the difference between the average American height of 175 cm and the 177 cm that has been measured, he would need to have made certain assumptions about the way heights are divided over the Russian army. Now people's heights are a product of a huge number of factors: genes, food, lifestyle, and who knows what else. It would therefore be strange if the people in Russia were of two very different heights with nothing in between. It is, in fact, reasonable to assume that heights are clustered around a smooth, roughly symmetrical, bell-shaped curve with just one peak lying more or less exactly where the average height lies. This is called a normal (also Gaussian) distribution. *Normal distributions* are found in a huge variety of measurements, including many of the ones discussed in Chapter 3. To assume normality is always a little hazardous, but our general had the benefit of inspecting the heights of Americans (see Fig. 5).

This data is roughly normally distributed, and this suggests that Russian heights might also be normally distributed. One other factor needs to be taken into account before a statistical test can be applied, namely the amount of variation in people's heights: if all Russians are approximately the same size then a small sample can go a long way, but if there is a lot of variation then more people need to be measured to get a good picture. To assess variation, statisticians use the concept of *standard deviation*, which is quite an intuitive concept: it is the average distance between the measurements in a sample and their average (i.e. the average value of the measurements themselves). You can think of this as the 'width' of the curve: the wider the curve, the higher the standard deviation. If the data are as depicted in Fig. 5 then the standard deviation of American heights is about 2 cm. (This is almost certainly an underestimation, but let's leave that aside.) Assume that the general— obsessed as he was with these matters—had read the literature on people's heights and learned that height variations are very similar across all countries; he took this to mean that the standard deviation among Russian heights was also 2 cm. Having made all these assumptions, everything was new in place for our general to perform a statistical test.

Essentially, a significance test would ask whether the Russian height measurement (177 cm, in our case) is *so* much higher than the American average (of 175 cm) that a height of at least that size would be unlikely to have been found by accident if the actual average of all Russian soldiers' heights was equal to the American average. In other words, the test measures the likelihood of jumping to an incorrect conclusion. The idea is to study the bell-shaped curve, and measure the size of that area under the graph which lies to the right of 177 cm, as a fraction of the area under the graph as a whole. Had one measured a greater height (say, of 178 cm), the area under the relevant part of the graph would have been smaller. Likewise, if the standard deviation had been smaller (i.e. the width of the bell had been smaller) then the area to the right of 177 cm would have been smaller than it is.

But even if all the general's assumptions were correct, a measurement of just one Russian soldier would not allow him to reach reliable conclusions about Russian soldiers in general.[8] If he had used just this one measurement of 177 cm the test would have told him that the likelihood of error is a shockingly high 15.9 per cent: given that the figure *without* any measurements would be about 50 per cent, this is little more than an educated guess. Suppose now that the general, aware of this issue, had measured nine Russians instead of just one, once again finding that their average size was 177 cm. Suppose he had used this to conclude that Russian soldiers are taller than American ones. This time, the likelihood of getting it wrong (given the general's assumptions concerning normality and standard deviation) was about 0.13 per cent. This percentage could be seen as measuring the rashness of his generalization: At 0.13 per cent, his guess is not as rash as it would have been after just one measurement. The details of the calculation will not concern us, because the basic idea is the same as in the case of a sample of one: the test estimates how likely we are to get it wrong when we make an empirical claim.

But now for the big question: is a rashness of 0.13 per cent something to lose sleep over? It would be nice if this question could be answered objectively, but it cannot, because our sleep is not something that statisticians have much to say about. No amount of calculation can tell us how much rashness we should accept. It depends on how much hinges on the truth of one's claims: if war and peace depend on it then a certain amount of care will be taken (one would hope). Different research communities use different significance thresholds. In some areas of psychology, for example, any error risk of less than 5 per cent is seen as acceptable; in medical science, where more tends to hinge on the outcome of a test, a more cautious significance level of 1 per cent is more commonly used. Either way, if the general found an average of 177 cm in a sample of nine Russians, then both of these thresholds suggest that the general had confirmed his claim, while if he had found it in only a tiny sample of one, the claim remained unconfirmed. With an intermediate sample size of four Russians it depends on whether he wanted to be as cautious as a medical scientist, or 'only' as cautious as an experimental psychologist.

Returning to the two meanings for the word 'significant' listed in the *New Penguin English Dictionary*, it is crucial to distinguish between the two, because a tiny difference may well be statistically significant: even the tiniest difference in average heights between your sample of Russians and the average of all Americans would become statistically significant if the sample were large enough and the standard deviation small enough. Presumably, a result of this kind would fail to be significant (i. e. important) in the other sense of the word. It is, in other words, quite possible for a result to be statistically significant but *insignificant* in any other sense.

Secondly, statistical significance is a gradable notion. The story about statistical significance resembles what we saw in connection with obesity: science gives us good reasons for adopting a particular measurement scale, but the reasons for adopting a particular set of *thresholds* are a good deal less principled. Why use percentages like 1 per cent and 5 per cent, instead of 1.66 per cent and 3.14 per cent, for example? Our liking for round numbers must surely have something to do with these choices (so if we had a different number system, we would probably have chosen different thresholds). Certain precisely specified thresholds have entrenched themselves in various research communities, but they are about as rational as throwing a big party when we turn 30: with equal justification, we could have post-poned the party until we turn 33.33, as a friend of mine did. Neverthe-less it is frowned upon to brag too much that a hypothesis was *almost* confirmed. In our quest for false clarity, confirmation is often treated as crisp rather than gradable. The consequences can be cruel: one medicine might be considered safe and another unsafe, based on only the smallest difference in experimental outcomes. But it is with significance as with democracy: it's the worst system imaginable—except for all the others.

With this, we are getting to the end of the first part of this book, whose main aim it was to explain the subtle role of vagueness in many of the concepts that we use, in everyday life and in science. In Part II we shall turn to linguistic and logical theories that have been proposed to make sense of vagueness.

Things to remember

◆ The fact that vagueness abounds in the presentation of mathematical results suggests that vagueness plays an important role in our thinking, even when the concepts about which we think are completely crisp.

◆ The notion of a 'small' number plays an important role in computing (in the form of maxint, which is the largest number that can be represented) and in a few less-studied areas of mathematics. Building a coherent mathematical theory of small numbers is difficult, because the sorites paradox looms large.

◆ Ordinary languages, such as English, assign a particular role to statements that involve round numbers, such as 'a thousand', 'a million'. Such statements have a tendency to be interpreted vaguely rather than crisply. This has unpleasant consequences for translation between languages when the languages in question use different metrics (e.g. when 'a yard', or 'a verst', is translated into a language that uses metres and kilometres only).

◆ Comparing the sizes of numbers is one thing, but comparing the sizes of functions (as computer scientists routinely do when assesing the run time of a program) is quite another. The standard way of doing this sacrifices practical applicability for mathematical cleanness.

◆ Statistical significance is a gradable concept. The thresholds that are commonly used to decide whether something is statistically significant are different for different research communities, and almost entirely arbitrary.

Part II

Theories of Vagueness

6

The Linguistics of Vagueness

We have seen that vagueness plays a huge role in human communication. Naturally, therefore, linguists have a lot to say about vagueness.[1] In this chapter we will highlight some of the main insights to come out of their work. Before focusing on vagueness itself, we shall introduce the methods that are used by modern linguists. These early parts of our story will centre around two very interesting—and very different—people: Noam Chomsky and Richard Montague.[2]

Chomsky's machine: Computing grammaticality

As someone who has done groundbreaking work in the mathematics of formal languages and in linguistics, while also writing numerous influential books on political topics, Noam Chomsky must surely rank as one of the most important thinkers of our time. In what follows, the spotlight is on Chomsky's linguistic research programme alone, as conceived in the 1960s and developed in later decades.

Chomsky, who has worked at the Massachusetts Institute of Technology for most of his life, was one of the first to envisage a formally rigorous description of the *syntax* of natural language. The word 'syntax' comes from the Greek verb *syntasso*, which means 'to organize, or to combine into an orderly whole'. The word was used in a military sense (involving the soldiers in an army) before—still in ancient times—it was applied to the words in a sentence. The study of syntax, or more elaborately syntactic structure, involves questions about the ways in which words form *patterns*, the most important of which is the sentence. Chomsky appears not to have had a strong interest in computers at the time, but I believe that his vision is best explained by talking in terms of what a computer might do with language.

At the heart of Chomsky's vision is the ideal of a machine that is able to tell whether something is a grammatical sentence (for example of English). You might picture a form on a page on the World Wide Web, into which you can type any string of characters. After a bit of computation, the system would tell you whether the sentence is a proper sentence or not:

Input: ... (some string of characters) ...
Output: This is/isn't a grammatical sentence of English.

Thus, if you type 'Only John eats apples', the machine says 'Yes, this is a grammatical sentence'; if you type 'John only apples eats', it says 'No, this is not a grammatical sentence.'

I am not aware of any web page where questions of this kind can be answered, but let us call this *Chomsky's machine* nevertheless. A more advanced version of this machine might even distinguish between different ways in which a grammatical sentence can fail to make sense. A sentence such as 'Colourless green ideas sleep furiously'—the example is Chomsky's—might obey all the grammar rules of English, but is non-sensical all the same, for instance because ideas cannot be green.

Chomsky held strong views. For instance, he believed that, underneath all the superficial differences between human languages, they are all fundamentally the same, springing from people's innate language ability. Chomsky's holy grail would be a version of his machine that is able to deal with every human language. Part of the ideal is to use just one machine, instead of separate machines for each language. This would have to be a *parametrized* machine, which works in subtly different ways

94

depending on the language involved, a bit like a spreadsheet many of whose figures can be affected if one variable is changed. Although we shall take no position on whether such a parametrized program is feasible, we shall act in Chomsky's spirit by focusing on a phenomenon—namely vagueness—that does seem to play the same role in every human language.

Furthermore, Chomsky was a syntax man. His domain is the patterns that words can form together: phrases, clauses, and sentences. He was, and probably still is, sceptical about the possibility of systematically describing the information conveyed by these patterns—their meaning, in other words. Vagueness did not interest him greatly, because vagueness is all about the meaning and function of language. Chomsky was certainly right in arguing that these domains are harder than syntax. But to exclude the very function of language (i.e. to communicate information) from linguistics is quite a limiting move, a bit like a doctor who refuses to cure any patients, limiting himself to diagnosis because he finds this easier. Time to let another giant enter the scene: Chomsky's fellow American Richard Montague.

Montague's machine: Computing meaning

In the previous section we have talked about Chomsky's work on grammaticality. Knowing whether something is a bona fide sentence is interesting, but what most of us are really interested in is the information that a sentence conveys. 'John eats only apples' and 'Only John eats apples', for example, are both impeccable English sentences, but this does not tell us whether they convey the same information—which, of course, they do not.

Putting this differently, what most people are interested in is the meaning of a sentence. But what *is* the meaning of a sentence? Researchers have pondered long over this question, and many different kinds of answers have been proposed. One kind of answer, however, has been particularly fruitful, and this answer is due in large part to Richard Montague.[3] Montague was a logician, philosopher, and linguist (as well as an accomplished musician) who worked at UCLA in California. Mystery surrounds his death—apparently his murder was never solved—but not his ideas, for his writing was highly precise, though often extremely terse. One of Montague's most famous papers is provocatively entitled 'English as a formal language': a formal language is the kind of precisely specified symbolism beloved of mathematicians and computer scientists. English, by contrast, is a naturally occurring language, and such languages have a reputation for messiness. The paper opens with the words:

> I reject the contention that an important theoretical difference exists between formal and natural languages. On the other hand, I do not regard as successful the formal treatments of natural languages attempted by certain contemporary linguists. (...) I regard the construction of a theory of truth (...) as the basic goal of serious syntax and semantics; and the developments emanating from the Massachusetts Institute of Technology offer little promise towards that end.

There can be little doubt that Chomsky was the main contemporary linguist Montague had in mind. What he meant by 'a theory of truth' I shall explain below: we shall see that the idea is a good deal more down to earth than it sounds.

Montague may have had as little interest in the construction of linguistic computer programs as Chomsky, yet Montague's ideas are best explained by taking a computational stance once again. Imagine a huge database containing all the facts that we are interested in at a given point in time. Now consider any sentence approved by Chomsky's machine. Suppose we had a second machine that was able, given any sentence of English, to determine whether the sentence is true or false. If the machine can perform this task reliably *regardless of how we fill the database,* and regardless of which English sentence we feed it, then in some sense, the machine 'understands' the meaning of all English

THE LINGUISTICS OF VAGUENESS

sentences. Montague's idea was that this task might be accomplished if the problem was approached in highly systematic fashion, starting from the words of the language, and working one's way upwards to larger and larger combinations of words, always 'computing' the meaning of a larger expression in terms of the meanings of its parts. This important idea is known as the *principle of compositionality*; its origins go back to the writings of the nineteenth-century German philosopher and logician Gottlob Frege.

To see the force of the principle of compositionality let us look at an actual computer program, which was inspired by Montague's ideas. The program is called TENDUM, a late representative of a series of so-called question-answering systems developed during the 1980s at Philips Electronics.[4] I use this as an example not because these systems are better than others, but because they illustrate Montague-style compositionality very directly and because I was fortunate enough to work on the development of one of these systems and, as a result, I am thoroughly familiar with them.

TENDUM was designed to give users information about international flight schedules, and it was able to answer questions such as 'Are there any direct flights from New York to Denver?' The fact that the sentence is in interrogative form (unlike 'There are direct flights from New York to Denver') introduces some problems of its own, but these are small enough that they can be disregarded here. TENDUM also allowed users to ask so-called wh- questions, such as 'Which flights are going from New York to Denver?', or indirect versions of the same question, for example 'Do you know which flights are going from New York to Denver?' Wh- questions are often more useful than Yes/No questions but, for convenience, we focus on Yes/No questions.

TENDUM's database contained basic information about long-distance travel, in a simple tabular form, e.g. there was a table much like the following (only much larger in both its dimensions), containing information about air travel:

DEPART	ARRIVE	DEP-TIME	FLIGHT
New York	Denver	Monday 9:00	NW204
New York	Denver	Tuesday 9:30	AL2002
New York	San Francisco	Monday 10:30	AM309
New York	San Francisco	Friday 15:30	AM4040
Denver	New York	Wednesday 20:00	NW704
Denver	New York	Thursday 20:00	AL2323
San Francisco	New York	Wednesday 7:00	AM3030
San Francisco	New York	Thursday 10:00	AM200

Another table contained information about ownership of airlines, for example:

FLIGHT	OWNER
NW204	Northwest
KL207	KLM
AL6767	Alitalia

Yet another said where each carrier had its company headquarters:

COMPANY	HEADQUARTERS
Northwest	San Francisco
KLM	Amsterdam
Alitalia	Milan

TENDUM analysed each question that was posed to it and converted the question into a computer-readable format that allowed it to access the database and determine the answer. For example, our question 'Are there any direct flights from New York to Denver?' would be converted into a formula of essentially the following form, using curly brackets to indicate collections of objects:

$$\text{Number}(\{\text{Flight}: \text{Depart}(\text{Flight}) = \text{NY} \ \& \ \text{Arrive}(\text{Flight}) = \text{Denver}\}) \neq 0$$

The formula says that the number of Flights that Depart from New York and Arrive in Denver does not equal 0. This statement, of course, comes out true given the database that we specified, because there are two such flights:

$$\text{Number}(\{\text{Flight}: \text{Depart}(\text{Flight}) = \text{NY} \ \& \ \text{Arrive}(\text{Flight}) = \text{Denver}\}) = 2$$

TENDUM computes the meaning of a sentence by associating a well-understood mathematical object with each part of the sentence. The simplest formula will be chosen that still allows you to compute correctly whether the sentence is true or false in a given case. Let us see how this works, working our way from the bottom upwards. First, what is a flight? Words can be tricky to define, but if all we want to do is compute the answers to our question then no full-fledged definition is needed: we can simplify drastically and stipulate that the word 'flight' 'means' the things in the set of Flights:

{NW204,AL2002,AM309,AM4040,NW704,AL2323,AM3030,AM200}

This listing is called the *extension* of the word 'flight'. Simple though the extension is, it suffices to answer such simple questions as 'Are there any flights (in your database)?' But that is not what we are after: we want to know about some smaller set of flights that go from New York to Denver. To compute the meaning of the expression 'flights from New York to Denver', we need to know the syntactic structure of that expression, so let us assume that we have a program that is able to add brackets to this phrase, as in

((flights from (New York)) to (Denver))

where the parentheses show which parts of the expression belong together. Then our next target is to compute the meaning of (flights from (New York)). Without going into the details of how this is computed—the key is to look up in a special dictionary that 'x is from y' can mean Depart(x) = y—it will not be surprising that this is translated as

(flights from (New York)) = {Flight: Depart(Flight) = NY}

after which an analogous procedure gives us

((flights from (New York)) to Denver) =
{Flight: Depart(Flight) = NY & Arrive(Flight) = Denver}

The essence of this procedure is to associate a semantic operation (i.e. an operation involving the meanings of words) with every way in which parts of a sentence can be joined together: with every syntactic operation, as we say. For example, the syntactic operation that combines 'flights' with 'from New York' to give 'flights from New York' is associated with the semantic operation that takes the set of Flights, whose extension was given above, and 'from New York', whose extension might be

{NW204,AL2002,T300,T400,AM309,AM4040}

(letting T300 and T400 denote trains, for example) and computes what these two sets have in common, namely

{Flight: Depart(Flight) = NY} = {NW204,AL2002,AM309,AM4040}

This, in a nutshell, is how TENDUM operates. Did TENDUM understand English? The first things to remark here is that the program understood a very small part of English at best, since there were many sentences that it could not make sense of at all, even in the area of air travel. For example, if you had asked '*Have you got any direct flights from New York to Denver for me?*' your sentence would not have been understood: TENDUM's grammar was able to analyse only a tiny fragment of English, and this limited its usefulness severely. Secondly, TENDUM contained only a small database, whose content did not exhaust all information about long-distance travel, let alone other topics. For reasons of this kind, programs such as TENDUM represent only the start of a long research programme, which is still far from finished.

It should be noted here that we have simplified Montague's idea considerably. A purist might even object that the procedure outlined above misses the point of Montague's greatest contribution entirely, which involves cases where the idea of identifying the meaning of an expression with its extension breaks down. This point is closely connected with another limitation: there is more to words like 'flight' than TENDUM captures. We shall return to some of these issues in later chapters.

For now, just note how awkwardly vague expressions fit into Montague's mould. Suppose we wanted to query the database about 'fast flights'. To do this, the database would have to hard-code what the words in these expressions mean, deciding for each flight whether it is fast or not. What do we do with boundary cases, involving flights that some would call fast and others slow? The system's decisions in such borderline cases might easily misrepresent the way in which a particular user understands it. Also, what counts as fast may change with time so it may be unwise to cast this in stone. We shall soon say more about these issues, which might be seen as posing a challenge to Montague's claim that there exists no 'important theoretical difference' between formal and natural languages.

The role of language corpora

Linguistics is a funny science, in that much of its wisdom can be attained from the comfort of an armchair. In this regard, traditional linguistics is unlike most other academic disciplines. Linguists can work in this way because, like any speaker of the language, the linguist carries with her, in her head, a rich repository of information about the language that she speaks. Through mere introspection, she can answer many of the questions that one might ask about her object of study. This truly is one of the delights of linguistics, just as it is of mathematics. But what if another linguist (or, heaven forbid, a lay person) disagrees with the linguist's answers? In mathematics, the response is to sit together and battle it out. This is possible because, ultimately, a mathematical result rests on proof. Linguistics is different because it concerns facts that could just as well have been different. Proof does not come into it. The string of characters reading 'John only apples eats' *could* have been proper English—It just isn't. If readers were to disagree with this claim, I would be unable to answer them from my armchair. Tangible evidence would be called for.

For reasons of this kind, linguists are turning to data that are more easily accessible to others than their private linguistic 'intuitions', and less dependent on the peculiarities of one particular speaker. Sometimes the data are the result of a controlled experiment involving human subjects; the subjects are offered some sort of stimulus (e.g. the experimenter asks a question) whereupon the subject's reponse is measured. In most cases, data take the form of publicly available repositories of 'real' language use, which are called language *corpora*. Ideally, corpora exist in electronic form, which enables some issues to be settled automatically. As a result of this development, linguistics is slowly turning into an empirical enterprise: one which rests on observation. The first electronically available corpora were limited in size. The Brown corpus, for example, which appeared as early as 1964, consisted of a number of texts which, collectively, contained 'only' about a million words (that's about six times this book).

A corpus that present-day students of English often use is the British National Corpus. The BNC contains texts, selected from newspapers and other publications of various kinds. Taken together, the texts in the BNC contain over 100 million words. Corpora, which can be searched by computer in increasingly sophisticated ways, are starting to influence the way linguists think about grammaticality. The classic view of grammaticality is as a *dichotomy*: Chomsky's imagined machine aimed at separating the strings that are grammatical from the ones that are not. But difficult questions can be raised. For example, taking our departure from a corpus-based approach, one wonders what to do if a particular type of expression does occur in one's corpus, but only rarely.

For example, suppose we find that split infinitives (e.g. 'to boldly go') occur much less frequently than their uninterrupted counterparts (e.g. 'to go boldly'), but still quite frequently? A possible response is to say that split infinitives are *somewhat* grammatical; one might even measure grammaticality using numbers between 0 and 1. Some linguists have argued that the whole linguistic edifice should be 'graded'. The machine envisaged by Chomsky should, in their view, be replaced by a more sophisticated machine that tells you, for any string of words, how grammatical it is:

> Input:...
> Output: The degree of grammaticality of this string is n.

This idea gives rise to many new questions. Does the frequency of a linguistic phenomenon tell us anything about the way it is represented in the mind of a speaker, for example? For example, are speakers able to estimate linguistic frequencies adequately, in reasonable correspondence with the frequencies that actually occur? The jury is still out on these questions, and they will not be pursued here any further.

For simplicity, we shall stick with the usual dichotomy between grammatical and ungrammatical strings in what follows. But this simplification does not remove vagueness from the agenda. For grammatical sentences may contain vague words, so the question arises how the meaning of such sentences may be computed, in other words, how Montague's paradigm might be applied to them. We shall focus on vague adjectives.

Vague adjectives

When we hear of the coastline of an unknown country, we come to it with certain expectations. We do not expect it to have some neat geometrical shape, like that of a straight line or the fragment of a perfect circle: we expect it to look a bit disorganized and capricious. It is similar with *adjectives*: when we hear that some human language (which we may never even have heard of) has an adjective that they use to talk about, say, the length of a man's beard, then we expect this adjective to be vague (i.e. admitting of boundary cases). Vagueness, in other words, is a universal of human language. The vagueness of adjectives has been studied more extensively than that of other types of words, so it is natural to start our survey with them.

First of all, what *is* an adjective?[5] Some words (such as 'tall', 'good', 'green') fulfil all the criteria for adjectivehood that have been proposed in the literature, including the fact that they can precede the noun (as in 'The green sofa') and follow it too (as in 'The sofa looks green'). But such words as 'tantamount' can only follow the noun, while such words as 'utter' can only precede it. You can't say 'The disaster was very utter', for example. Linguists deal with this diversity by distinguishing between typical adjectives, which can be used attributively and predicatively, and peripheral ones, which can be used in only one of the two ways. Let us use the above-mentioned BNC corpus to find out which adjectives are used most frequently. It turns out that these are the top ten: last (140,063 occurrences), other (135,185), new (115,523), good (100,652), old (66,999), great (64,369), high (52,703), small (51,626), different (48,373), large (47,185). Let us disregard the peripheral ones in the list (i.e. 'last' and 'other') and turn to vagueness. Which of the top eight adjectives would we want to call vague? One way to frame this question is by asking which of these eight allow boundary cases, in which case the answer is, probably, all of them. Another way to answer the question is proposed by Quirk et al. (1972), who suggest adding the following two questions to the ones mentioned above: can the word be modified by the intensifier 'very'? And, does the word allow comparisons (as in 'Andrew is more...than Harry', or 'Andrew is ...-er than Harry')? If the answer is affirmative in both cases, the adjective is usually called a *degree adjective*. Checking all eight 'central' adjectives in the list, it appears once again that they all comply.

Whichever way we look, the most frequent adjectives tend to be vague. Rather than discuss why this might be—are these words frequent because they are vague, or are they vague because they are frequent?—let us see what havoc is caused by their vagueness.

The meaning of adjectives

Meaning doesn't have to be a nebulous concept. To paraphrase the words of the famous philosopher David Lewis, if you want to know the meaning of an expression, you have to find out what its meaning *does*, then find a mathematical construction that does just that. As it is often put more briefly: *meaning is what meaning does*. This thoroughly 'Montagovian' perspective will be useful when thinking about the meaning of vague adjectives. The idea is to ask how adjectives contribute to the meaning of the larger expressions of which they can form a part.

Let us first consider ordinary, crisp adjectives, such as the word 'pregnant'. What does 'pregnant' mean? In the spirit of the previous section, where the TENDUM system was described, we might answer that its meaning can be approximated, given a certain domain of objects, by identifying it with its *extension*: the set of individuals that are pregnant. The way in which this meaning contributes to the meaning of a larger expression of the form 'pregnant [Noun]' is through forming the intersection between two sets (i.e. the set of objects belonging to both sets). Consider, for instance, the expression 'pregnant grandmother'. We compute its meaning by forming the intersection between the set of grandmothers and the set of pregnant individuals. This gives us a starting point for computing the meaning of even larger expressions, including complete sentences such as 'All pregnant grandmothers in Aberdeen live in Torry'. Adjectives such as 'pregnant' are called extensional because (to stay with our example) you have to know only who the grandmothers are, who lives in Aberdeen, and who is pregnant to know who the pregnant grandmothers in Aberdeen are. To judge the truth of the sentence, the only other thing you need to know is who lives in Torry. All you need are extensions of words.

But some adjectives operate differently. Consider the adjective 'future'. What is the set of all future grandmothers? Whatever it is, it probably does not consist merely of grandmothers! To say that a future grandmother is, by definition, something that is both future and a grandmother would be dead wrong. Something about the word 'future' (likewise 'former', 'would-be', 'certified', and so on) causes it to have a different effect on the meaning of the expressions of which it is a part from the word 'pregnant'. This phenomenon—which is at the core of Montague's machine—is called *intensionality* because, to compute the effect of these adjectives, it is not sufficient to take the extension of the words in its vicinity into account. In a nutshell, to compute what the set of would-be grandmothers is, it is insufficient to know who the (present) grandmothers are: you have to know what is called the *intension* (written with an 's') of the word 'grandmother', essentially an oracle that tells you, for each moment in time, who the grandmothers are at that point in time. Variations on this theme can be employed to compute the set of future grandmothers. Far from being limited to issues of time, intensionality can stem from a range of sources.

In this book, we are often concerned with another class of adjectives, which resist a simple extensional treatment for different reasons from intensional ones such as 'future'. Consider the adjective 'young'. What's a young grandmother? The simple extensional treatment says 'find everything that is young, then find the grandmothers, then find the things that these two sets have in common'. It is easy to see that this recipe will not work. One can phrase the matter in terms of expectations: if you expect to see a grandmother, you expect to see someone who is fairly old. Language has a way of compensating for such expectations, so that when we call someone a young grandmother, we are saying that she is young for a grandmother.

What this means is that degree adjectives do not always have the same extension: their extension depends on the words surrounding them, and perhaps even the context in which they are uttered. It is useful to ponder why language plays this trick on us. The reason is not difficult to see: If

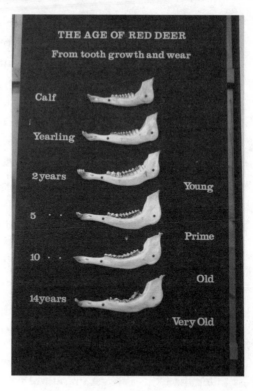

FIG. 6 Young and old in
the context of deer

the word 'young' always had the same extension then this would severely limit its use. Suppose, for instance, words such as young and old were specifically coined with toddlers in mind. Then, surely, all grandmothers would count as old, making 'old' and 'young' useless for discriminating between different grandmothers. When I was a child, I imagined language being invented by a committee of scientists, who made wise decisions about word meanings and the like. Clearly if the scientists were any good, they would have made words that measure things dependent on context, in the way that 'old' in 'old X' is dependent on X, and this is exactly what was done. The alternative would have meant having to coin countless different words, each of them meaning 'young', but applied to a different kind of thing.

107

The idea that the meaning of degree adjectives tends to be dependent on context is uncontroversial. Unfortunately, this is where the agreement ends. For if we ask *how* context influences meaning, then different answers have been forthcoming. When we describe an object or a person, it is often far from easy to determine what it is that we are comparing them with. When we see 4-year-old Johnnie and exclaim 'Isn't he tall', it is difficult to say what exactly we are comparing him with. This is analogous to our discussion of relative poverty, in Chapter 3, where it proved difficult to determine what it means to call a person 'poor', because you could compare them with different sets of people. What is clear is that, in expressions of the form 'the tall so-and-so', the class of all the so-and-sos often plays a particular role. Linguists have attempted to capture this role in different ways. For example:

tall X = those X that are taller than the average X; or
tall X = those X that are taller than most X.

On the first definition, a tall grandmother would be a grandmother who is taller than the average grandmother, for instance. Whether any of these definitions is at all accurate seems doubtful. To start with, the fact that both definitions turn such expressions as 'tall basketball player' into crisp ones should, in my view, raise suspicion: the definitions above deny that words such as 'tall' are vague at all, because they do not allow any borderline cases. But let us leave logical issues aside for now and concentrate on more narrowly linguistic ones.

Presumably, if we see five grandmothers of very different heights, we can talk about 'the tall one' to designate the tallest of the five, but equally of 'the two tall ones' to designate the tallest *two*. But this means that the word 'tall' must have had different meanings in both expressions: one definition of the kind given above could never account for both usages. Observations of this kind are starting to suggest that speakers are, to some extent, free to choose what they mean by a word such as 'tall'. In the words of Alice Kyburg and Michael Morreau, 'Just as a home handyman can fit an adjustable wrench to a nut (...) a speaker can adjust the extension of a vague expression to suit his needs.'[6]

To rub in this point a bit more, consider another type of vague expression, based on words such as 'many' and 'few', which linguists call determiners or quantifiers. The Scottish psycholinguists Linda Moxey and Tony Sanford performed two different experiments involving these expressions (Moxey and Sanford 2000). In the first experiment, they asked people to say how adequately a given numerical quantity would be described by a sentence of the form 'many A are B'. Predictably, the higher the percentage of Bs among the As, the more correct subjects found these expressions (and the less correct expressions of the form 'few A are B').

The surprise lies in their second experiment: when each subject was asked to make *just one* judgement instead of several, there was no longer the expected correlation between the percentage of Bs and the average degree of correctness attributed to a given sentence pattern. The authors concluded that the outcome of the first experiment was an artefact, caused by the fact that people were given a large set of highly similar tasks, which allowed them to make comparisons they would not normally make. Care must be taken when generalizing these results to adjectives, but Moxey and Sanford's research suggests that people's 'real' understanding of vague expression is governed not by objective percentages, but by what the speaker *expected* before being confronted with the facts (and also what the speaker believed the hearer to expect). Applied to adjectives, this suggests a definition along the following lines:

Tall X = those elements of X that are taller than the speaker
   expected the elements of X to be

It appears, incidentally, that context is not just a linguistic concept, but a deeply ingrained psychological mechanism. We saw how context played a role in our discussion of relative poverty, for example. Or consider the way in which we talk about the people around us. A couple with two children might think of one of the two as healthy (or cheerful, or bright) and the other as sickly (or gloomy, or dim), even if the differences between their offspring are small in comparison to the population as a whole. When these parents describe one child as, say, the cheerful one, then what started out as a mere comparison may well be experienced as an absolute: to the parents—and perhaps to the children as well—it feels as if the child is cheerful *full stop*. Context gives them a pretext for talking as if other children did not matter.

Vagueness and ambiguity

It is time to bite the bullet and ask: how does linguistic vagueness differ from other ways in which language can be unclear?[7]

To illustrate what I'm talking about, let us return to the TENDUM computer programme discussed earlier in this chapter, when we showed how the principle of compositionality can be cashed out by a language-interpreting machine. (Once again, a computational angle will make it easier to understand what goes on in language.) So far, we have focused on such queries as 'Are there any flights from A to B?' But TENDUM had to work hard to interpret queries that contain words that can be defined in different ways. The word 'American' is a good example, as in 'Are there any American flights from A to B?' The problem is that 'American' can be interpreted different ways. To say that x is American could mean each of the following things:

(a) country(birthplace(x)) = USA
   (The birthplace of x is in the USA)
(b) manufacturing-country(x) = USA
   (x was made in the USA)
(c) country(depart(x)) = USA
   (The city from which x departs is in the USA)
(d) country(headquarters(owner(x))) = USA
   (x's owner has its headquarters in the USA)

(e) owner(x) = American Airlines
     (*x is owned by the company called American Airlines*)

Since flights do not have birthplaces, interpretation (a) can be ruled out. TENDUM does this by a selection device called *typing*: the function 'birthplace' in (a), for example, is constructed in such a way that it can be applied only to people. Similarly, the function 'manufacturing-country' in (b) can be applied only to actual aircraft. The functions 'depart' and 'owner', however, can be applied to flights. If the programme attempts to apply the formula (a) to a flight f then a *type conflict* arises: country (birthplace(f)) = USA simply does not make sense given that f is a flight and flights do not have birthplaces. The same is true for formula (b), which requires an aeroplane rather than a flight. The programme keeps only those interpretations that do not give rise to a type conflict: (c), (d), and (e) in this case. The first of these three, for example, will cause American(f) to mean country(depart(f)) = USA, meaning 'The city from which f departs lies in the USA', in ordinary English.

Using types to get rid of nonsensical interpretations (involving the birthplace of a flight, for example) is very useful. Unfortunately, however, in the case discussed above there are still three interpretations left, each of which makes perfect sense in combination with a flight. Each of them gives rise to a separate question, with potentially different answers. This is a problem, since the computer may accidentally choose to answer a question that the user has not asked, potentially causing misinformation. The root of the problem, in this particular case, lies with the word 'American', but words are not the only cause of this phenomenon, which is generally known as *ambiguity*. My aim here is not to explain how computer scientists deal with ambiguity—a topic on which there exists a very substantial literature—but to understand as well as possible how ambiguity differs from vagueness.

At one level, vagueness and ambiguity are manifestations of the same phenomenon. They are the kinds of thing ordinary languages are full of, unlike programming languages and the formal languages of mathematicians, for example. Just as the vague adjective 'tall' has different thresholds depending on what it applies to (people, buildings, etc.), the adjective 'American' denotes differents kinds of 'American' depending on what it applies to (people, aircraft, and so on). In both cases, the phenomenon can be explained from a principle of efficiency: a child needs to learn a word such as 'American' (or a word such as 'tall') only once, after which she will be able to apply it to a variety of different situations. This is obviously a huge advantage.

But the price we pay for this efficiency is a risk of misunderstanding. TENDUM was often unable to settle for a unique interpretation of an ambiguous word such as 'American', for example. For, as we have seen, to know that the word applies to a flight helps to narrow down the set of possible interpretations of the word, yet several interpretations remain. The same is true for vague words. Suppose we are at a basketball match between two teams in a lower division, none of whose players are particularly tall. Suppose we refer to a player as short. *First*, what is the context that should help us to understand what is meant by 'short'? Does it consist of just the players on this particular team? Basketball players in general? People in general? This type of quandary is far from rare: wherever context affects interpretation, occasional misunderstandings are unavoidable, because it will often be unclear what the most relevant context is. *Secondly*, even if we have settled on a particular context, this does not pin down interpretation standards completely, because we need to decide whether 'short' means shorter than average, shorter than most, or something else again. Ambiguity and vagueness are similar in these important respects.

But there are differences as well. One difference between vagueness and ambiguity is that, whereas there is always something highly systematic and predictable about vagueness, ambiguity can sometimes be arbitrary: whereas the first four readings of 'American' are exactly analogous for the word 'Danish', or 'Moroccan', for example, this is not the case for the fifth interpretation, which involved the company American Airlines. Or consider an ambiguous English word such as 'pitcher': why the same word can denote a container for liquids as well as a baseball player is anybody's guess. The two interpretations seem completely unrelated. Language is partly systematic, but partly messy as well, resulting from accumulated accidents of history.

Another difference is that ambiguity arises, far more often than vagueness, from grammar. Vagueness resides mostly in words.[8] Ambiguity, by contrast, can arise from interactions between grammatical constructions each of which is entirely harmless by itself. The joining of two nouns into a new, conjoined noun ('women and children') normally does not produce ambiguity, nor does the joining of an adjective to a noun ('poor men'). But when both phenomena occur together, ambiguity can spring from nowhere, for example in the words 'poor women and children', where it is not clear whether the children or only the women are said to be poor. Syntactic ambiguities of this kind lurk everywhere, threatening to render an otherwise transparent text unclear. They are as good an example of the *emergent phenomena* beloved of chaos theorists as you are likely to see.

But the greatest difference between ambiguity and vagueness is a difference of quantity. The TENDUM system, for example, was able to get away with only a few readings for words such as 'American'. Given the relative simplicity of TENDUM's air traffic domain, any given speaker could, realistically, have only one of a few readings in mind. Compare this with the infinite number of possible ways in which words such as 'fast', 'tall', or 'small' could be intended and the difference is immediately clear. Also, each of the readings of 'American' is nicely, crisply, different from the other, whereas the different readings of 'tall' flow over into each other imperceptibly with the gradualness known from the sorites paradox. Indeed, ambiguity does not give rise to paradoxes like sorites. It is for these reasons that this book focuses on vagueness, leaving ambiguity more or less aside.

Having listed these differences, the question comes up as to how the two phenomena can be told apart. Suppose you were studying English as a foreign language, and you were to hit upon a new word. How would you decide whether the word is vague or ambiguous? Linguists have

THEORIES OF VAGUENESS

invented a test to help you. Let's see how it works, focusing on some ideas associated with the name of George Lakoff.

Consider the *ambiguous* word 'American'. Suppose I say 'John took an American flight, and Claire too'. Given what was said above, this sentence can have three bona fide interpretations: one for each applicable interpretation of 'American', once in connection with John, and once in connection with Claire. What is *not* possible is to choose one interpretation for John's flight (let's say interpretation (c) above) and a different one (say (d)) for Claire's. A *vague* word such as 'tall' appears to work differently: If I say 'John is tall and Claire too' then John and Claire can have different heights. The idea of the test is to say that words that work like 'tall' are vague, while words that work like 'American' are ambiguous. Similar tests rely on words such as 'all': If I say that all American flights are delayed, I cannot be talking about all the flights that count as American using criterion (c) *and* about all the ones that count as American using criteria (d) and (e); I must be focusing on one interpretation all the time. By contrast, if I say that all tall people are evil, I may very well be talking about people of different heights. Once again, the idea is to call a word vague if it behaves like 'tall' and ambiguous if it behaves like 'American'.

These generally accepted linguistic tests seem broadly on the right track: the idea is that ambiguity involves different interpretations while vagueness involves different ways in which a given interpretation can be made precise. Yet, I must admit not always feeling confident using the tests. Is it really possible to use the word 'tall' as outlined in the previous paragraph, while using different *thresholds* of tallness for the different things involved? Surely, in the example involving two people's heights, I must be judging John and Claire by the same yardstick? Suppose I was confused enough to want to talk about tall people and tall buildings in one breath, but using the very different thresholds that are commonly associated with people and buildings respectively. One couldn't say 'John is tall, and that building too', without humorous intent. Perhaps it's safer not to rely on these tests, and to use our earlier observations as the

114

dividing line between ambiguity and vagueness instead: if different interpretations arise from the fact that thresholds can be chosen at a huge number of different levels, we speak of vagueness; in other cases we speak of ambiguity.

Vagueness and ambiguity can also go together, causing even greater havoc jointly than in isolation. This happens, for example, when a vague word is used in a non-literal sense. Take a word such as 'big'. In addition to denoting physical size, it can also be applied to importance ('a big issue') or popularity ('big in Japan'), for example. On each of these interpretations, the word is also vague. Similar things hold for the word 'great'. It is extremely common for literal interpretations to be coupled with metaphorical ones, and some people have even argued that this phenomenon is the greatest obstacle preventing computer programs from genuinely understanding language. A lot of linguistic work is now going into trying to discover these mechanisms. All the available evidence suggests that the processes involved are highly general, and that they apply to crisp as well as vague expressions. In this book, we shall therefore leave these issues aside.

Lack of specificity

Suppose I glance into a university classroom and, based on a quick estimate, I tell you that more than a quarter of the students are female; perhaps I'm saying this in surprise, since it's unusual in computing science. Am I being vague?

If my statement is taken literally there is nothing vague about it. True, I am not as *specific* as I might have been: '30 per cent' would have been slightly more specific, for example. But none of this makes my utterance vague in the sense of this book, since my utterance allows no borderline cases: if more than 25 per cent of the students were female I'm right, otherwise I'm wrong. The difference between vagueness and lack of specificity is quite important. For even though the two phenomena have much in common, vagueness is much more difficult to handle than lack of specificity, as we shall see in the next few chapters.

But a subtler picture emerges if we allow ourselves to read between the lines. Consider the same situation, which has me peeking into a classroom. If I say that more than a quarter of the students are female then, normally, you're entitled to infer that their number does not exceed 25 per cent *by very much*: you'd be surprised to learn there were 80 per cent, and might even claim that I had misled you. For if I knew that there were that many, why didn't I make a stronger statement? Normally, to say that more than x per cent of people have a certain property means that this is the best lower bound that I am able to offer. The actual percentage may be higher, in other words, but not by very much. How much exactly is unclear. Because of this mechanism, unspecific utterances have a tendency to be interpreted as vague.

The principle that was just invoked is one of the most important ones to come out of modern linguistics so far. It is known as the *maxim of quantity* and is sometimes formulated as 'express yourself as strongly as your information allows'. Obvious though it may sound, its inventor Paul Grice demonstrated that this maxim forms a part of a sophisticated code of conduct that governs all human communication. This code does allow certain types of violations, but each violation bears information too. Suppose you were a seasoned observer of politics, whose judgement carries a lot of weight. A journalist has asked you what you think of a certain Mr X's suitability as a prime minister of Britain. Suppose you responded saying 'Mr X is basically a decent man, who has been an excellent chancellor'; at this point you shut up and take a sip of your tea. You would have been saying nothing but good about Mr X, yet your judgement would have been devastating because of the word 'basically', and because of all the prime-ministerial qualities you *failed* to mention. If Mr X has all these qualities (leadership, international stature, voter appeal), why didn't you say so?

Prototypes

Linguists come in different flavours. Some, like Chomsky and Montague, regard language as a mathematical object. Others, known as psycholinguists, study language primarily as a product of the human mind. For now, let us focus on one area in which psycholinguists have been active: they have found that people associate nouns with different types of object to different degrees, and this has given rise to an an area of study called *prototype theory*, which is associated with the name of the psychologist Eleanor Rosch. Consider the word 'bird', for example. A sparrow is, somehow, a more typical bird than a condor; a penguin is less typical. In an experiment with human participants, a picture of a sparrow, for example, is more rapidly recognized as a bird than a picture of a penguin. Moreover, there exists remarkable agreement among participants about the degree to which a given referent is a (prototypical) bird.

The findings of prototype theory have been confirmed in a large and varied number of experiments. It would be nice if regular linguistic theories, of the kind discussed earlier in this chapter, were able to take these facts on board, by representing the degree to which a word is true of a thing. In the case of words such as 'tall' this seems unproblematic, but it is harder in other cases. The problem is that people are only too willing to go the extra mile and tell you to what different degree, say, 3 and 23 are prime numbers, with baffling degrees of agreement, as Steven Pinker points out in his book *How the Mind Works*. Surely this does not mean that 3 and 23 are prime numbers to different degrees. So what *do* subjects' answers mean? Or consider Pinker's example of the word 'grandmother'. On the one hand, this word has a completely crisp definition: a grandmother is the mother of a parent, and nothing else. Yet a grey, soup-dispensing 70-year-old turns out to be much more stereotypical than a marathon-running 40-year-old whose daughter just happens to have had her first baby. As Pinker notes, crisp and fuzzy ideas about grandmothers

117

exist side by side in our heads, and the psychological implications of this cohabitation are far from clear at this stage.

Yet prototype theory is both interesting and potentially useful. In computing science, for example, prototypes have given rise to a new perspective on what it means for one group of things to be equal to (or part of) another. The resulting theory, a refinement of set theory developed by Tony Cohn and others at the University of Leeds, is known as 'egg-yolk hierarchy' (e.g Lehmann and Cohn 1994). Consider a concept such as 'employee'. Clear though it may seem, in the cut and thrust of real applications this concept turns out to possess boundary cases. It is unclear, for example, whether the head of a company counts as an employee, and the same is true of a lawyer who spends only a small part of his time working for the company. Such boundary cases can cause problems when computer programs are linked up with each other, because one program might, for example, include the head as an employee but not its lawyers, while the other program might have decided matters the other way round.

Egg-yolk theory takes the position that it is wise to allow a few mismatches between programs. To ensure that not too many errors are made, and not too crucial ones, Cohn and his colleagues advocate what they call, amusingly, the importance of doing things wrong: 'What is needed is a way to do most things possibly wrong, but probably not wholly wrong.' They achieve this by taking a leaf out of the prototype theorists' book, focusing on the *prototypical* members of a group. When a computer program is constructed, the programmer is asked to specify, among other things, what the prototypical members of its main concepts are. (Presumably, in the case of the concept 'employee', these are ordinary, run-of-the-mill employees who work for the company full time.) These prototypical cases form the 'yolk' of the egg. To check whether two concepts can be equated without too much risk, it often suffices to focus on the yolk: if their yolks are equal, it is safe to regard the two concepts (i.e. the two eggs) as equal. Some errors may result from any remaining differences between the concepts, but this is something egg-yolk theorists accept.

Comparatives

Degree adjectives come in three forms: the base form ('tall'), the comparative ('taller'), and the superlative ('tallest'). These forms are closely related to each other, and the relation between the base form and the comparative is of particular importance as we shall see in the next chapter.[9] For this reason, let us briefly dwell on the grammar of comparatives. As it happens, this is a difficult topic, because many sentences seem to lie in the grey area where grammaticality is doubtful. 'John is taller than Bill' is fine, but how about 'John is more heavy than tall'? Confident answers are difficult to give, yet some important issues have arisen from questions of this kind.

When analysing the meaning of sentences involving comparative expressions, such as 'shorter than', 'too heavy to...', 'so old that...', it is tempting to make use of *degrees* in one way or other. If an object is too heavy to be safely transported in a lift, for example, then the object is heavy to a certain degree (i.e. it has a certain weight), such that this degree surpasses some maximum value above which transport is unsafe. Similarly, if John is heavier than he is tall, then presumably he is heavy to some degree d_1, and tall to some degree d_2, where d_1 surpasses d_2.

After a number of years in which comparatives have been extensively studied, many linguists now favour an analysis of vague expressions in which degrees play a central role. 'Venus is brighter than Mars', for instance, means that the degree to which Venus is bright exceeds the degree to which Mars is bright. Even such a simple expression as 'John is tall' is now viewed as involving a relation between John's and some other degree of tallness: John is tall to some degree d, such that d exceeds some standard s. The standard s is not explicitly mentioned in the sentence, yet the analysis pretends that it is (as it were). The problem is determining how high s is. The obvious place to look for the answer is, once again, the context of the sentence by which we mean the previous sentence, or the conversational setting in which the sentence was uttered. If it was during a basketball match then *s* may be assumed to be the degree of tallness that corresponds with an average-sized basketball player. This analysis—which I can sketch only in its barest outlines—is typical for modern linguistics in its attempt to find a general pattern that covers very different sentences.

Perhaps the hardest question associated with comparatives is what to make of degrees. Suppose degrees can be pictured as points on a scale. Are they points on one and the same scale, or are some things incomparable with each other? Think of a would-be sentence such as 'John is heavier than he is nice', for example, where comparability is problematic. Questions of much the same kind will rear their ugly heads in Chapter 9, when we discuss fuzzy logic and other logical theories which make the notion of a degree their linchpin.

Packaging what we say: Hedging

> Ramón Rudd's room was huge (…) with tinted windows (…) and a desk the size of a tennis court and Ramón Rudd clinging to the far end of it like a very small rat clinging to a very big raft. (John Le Carré, *The Tailor of Panama*)

So far, we have kept the linguistic story of vagueness pure and simple. But vagueness can interact with other things that happen when we communicate: irony is a case in point, and so are exaggerations and metaphors, to name but a few.[10] As for exaggerations, a desk 'the size of a tennis court' may be large, but presumably not *that* large! So although Le Carré appeared to say something precise about the desk, in fact he did not. Similarly, if I say metaphorically of a surgeon that she is a butcher, I am not explicit about the exact nature of the similarity, or about its degree. Or finally, to use a classic example, suppose a company director says to his secretary 'Would you mind checking this letter for typos?' By putting this as a question—and a very polite one at that—he has chosen to leave it a bit vague whether this is essentially an order (at one end of the scale), or more like a suggestion (at the other end). Language has many ways to wrap a layer of soft padding around the things we say, and vagueness is implicated in many of them.

One kind of padding is important enough to dwell on, namely *hedging*. In keeping with the spirit of the preceding chapters, let us explore briefly what happens when hedging is used in the presentation of highly tangible information, for example when scientific results are communicated.

Scientists will seldom say that an experiment 'proves' something; they are more likely to say that their data 'appear to confirm the hypothesis that so-and-so', or words to that effect. Expressions such as 'appear to' are *hedges*. Hedges are not just words that you add to a statement to show that you've been to university. Hedges express uncertainty.[11] And since uncertainty comes in degrees, hedged statements tend to be vague as well. If you say that something appears to be the case, for example, you are not saying how much doubt surrounds this appearance. Admittedly, the uncertainty suggested by hedges is not always genuine: if a British policeman tells you 'I'm afraid I may have to handcuff you', resistance is futile. But let us disregard politeness and other, less pleasant forms of insincerity, focusing on the genuine article.

Every linguist has his own theory of what a hedge is, and this makes hedges difficult to count. One way to cut this knot is to focus on a small group of clear cases, such as the expressions 'may' and 'might', which are hedges in just about everyone's book. In an extensive study by the linguists William Grabe and Robert Kaplan, which focuses on these two words, some interesting patterns emerged when hedges were compared in various types of texts. (Such a narrow strategy is not without problems, but let us disregard these here.) It turned out that hedges occurred more than ten times as often in science texts than in business reports and newspaper editorials, and that they occurred hardly at all in narrative texts.

The rationale behind this finding is probably that true precision often requires admitting where it is that you are uncertain or imprecise. The popular view of science tends to depict it, quite incorrectly, as a bastion of certainties. This misperception makes the communication of science even more hazardous than it would otherwise be. Consider a climate scientist whose models predict a global warming effect of 5 degrees Celsius in 100 years' time, with a certainty of 70 per cent for example. If she tells a journalist about her findings, these may be reported as saying that 'the earth will heat up by 5 degrees in 100 years', thereby overstating her case. To prevent such misinformation, our scientist may emphasize her uncertainties. But if the journalist reports the predictions faithfully, then the resulting news report ('climate may warm by 5 degrees') is not likely to make much impact. It might even be attacked for being speculative or 'just a theory'. This conundrum is probably one of the reasons why warnings on global warming have been neglected for so long.

A more fully documented example is the run-up to the 2003 invasion of Saddam Hussein's Iraq. The British government, in a document that came to be known as 'the dodgy dossier', decided to remove certain hedges from classified documents. The original documents were careful assessments by the intelligence community concerning Saddam Hussein's weapons of mass destruction. To focus on one notorious example, early documents by the Joint Intelligence Committee said that 'Intelligence (...) *indicates* that (...) chemical and biological munitions *could* be (...) ready for firing within 45 minutes'. Later versions, aimed at parliamentarians and the general public, rephrased this, much more bluntly, as 'The Iraqi military are able to deploy these weapons within 45 minutes of a decision to do so'. This claim ended up playing a key role in Parliament's deliberations. Later research indicates that the original assessment had reached the intelligence community through a chain of three sources, one of which was considered less than reliable. In other words, the assessment was based on hearsay, so the Joint Intelligence Committee had been absolutely right to be cautious. By leaving out crucial hedges, politicians made sure that hearsay was presented as the unadultered truth. By thus 'sexing up' (as it came to be known) crucial information, the Blair government may well have caused the British Parliament to acquiesce in a military invasion that it would otherwise have condemned.[12]

Future work

Perhaps the largest remaining linguistic challenge relating to vagueness, which is strongly related to some of the logical issues that will be discussed in the next few chapters, is to obtain a better understanding of the interactions between vague expressions and the context in which they are uttered. Context influences the interpretation of vague words, but as the linguist Chris Barker has stressed, when we utter a vague expression, this action also changes the context. One way to think about the situation is in terms of a comparison: by saying that x is large, I am saying that x is above my threshold for largeness: $x > threshold$. If you knew my threshold, this tells you something about x, but if you knew x...then it tells you something about my threshold.

Here is an example of the latter situation: Suppose you are a student in a cooking class and the chef, after preparing a curry and letting you taste the result, says: 'This is a mild sauce.' The chef is *not* saying this to inform you how spicy the sauce is (since, having tasted it, you know this!) but to teach you what counts as mild in the case of this type of sauce.

Some utterances may have precisely this function: to tell you what the speaker's norms or thresholds are. But sometimes the situation is subtler: suppose I use the phrase 'the three large dogs' to point out some animals to you. Typically, I will be doing this for a practical purpose, perhaps to tell you which animals in a kennel I have just bought. Yet, my utterance is affecting the context as well. For example, it would now be inconsistent for me to refer to one of the three as 'the small dog' in the next sentence. Dogs that are even larger must certainly count as large. My utterance might even discourage you from calling another dog 'small' unless that dog is *considerably* smaller than the ones I just referred to. The development over time of the thresholds relevant to vague expressions is little understood at the moment and crying out for more research.

Things to remember

◆ Richard Montague's work gave linguists a framework within which it was possible to give systematic descriptions of the *meaning* as well as the form of English sentences (or those of any other natural language, for that matter). The framework built on work by grammarians such as Noam Chomsky, whose work focused exclusively on the *form* of sentences.

◆ Natural languages pose several challenges to Montague's framework. Vagueness is one of the most important of these challenges. Another is the related yet different phenomenon of ambiguity. Vagueness and ambiguity allow speakers of natural langages to say things briefly using a limited vocabulary, but these efficiencies come at a considerable cost: the risk that the hearer may fail to understand what information the speaker tried to convey.

◆ Vagueness should also be distinguished from other ways in which information can lack specificity: to say that '*At least one* student at this university passed this exam', for example, is unspecific because the number of students is not given. It is not vague, however, because the sentence offers a clear criterion for determining whether it is true. This is because 'at least one' does not have boundary cases.

◆ The risk of misunderstandings caused by vague and ambiguous expressions can often be diminished by the context in which these expressions occur. For example, a vague word can be clarified (to an extent) by the words around it. A 'short basketball player', for example, is someone who is short *for a basketball player*. It is important to note, however, that the resulting expression is still vague, because there exist people who are borderline cases of a 'short basketball player'. Thus, context reduces vagueness but seldom removes it entirely.

- ◆ Noam Chomsky believed that all natural languages obey essentially the same rules. At the time of writing it is unclear whether this is true for the syntactic structure of these languages, on which Chomsky focused his attention. But for many other aspects of language and communication, Chomsky must surely be right. It would, for example, be surprising if there was a language in which no vague adjectives existed, in which no sorites paradox could be formulated, or in which context did not play the same disambiguating role as in English.

- ◆ Hedges such as 'probably', 'apparently', and 'maybe' are essential indicators of uncertainty. Whoever removes them from a message does this at their own peril—and sometimes at other people's peril as well.

7

Reasoning with Vague Information

Deduction is usually associated with the manipulation of exact information. If I tell you that 1,000 identical sweets were divided as fairly as possible between two children, you would deduce that both children received 500 sweets each. But what if we use vague words instead of numbers?

If I tell you that a 'very large' number of sweets were divided fairly between a 'very small' number of children, there is no way you can work out how many sweets each child got. Still, there is room for deduction. You know that no child received in excess of 1 sweet more than any other child, or else distribution could have been fairer. And since there were a very large number of sweets and only a very small number of children, presumably no child was left empty-handed; you might even surmise that every child must have received a 'substantial' number of sweets. Conan Doyle's Sherlock Holmes loved arguments of this kind: some of his stories even hinged on them.[1]

Deductions involving vague concepts can take many different forms. Some, for example, exploit the way in which vague expressions rely on the words around them for their interpretation: their dependence on context, in other words. Suppose we are talking about the processing power of a computer, and we accept that the computer in your office is 'a powerful PC'. Suppose laptops are less powerful, on average, than PCs, yet your own private laptop happens to be at least as powerful as the computer in your office. Surely, it then follows that it must be 'a powerful laptop'. Words such as 'powerful' may be vague and context-dependent, but this does not mean we cannot reason with them. This is just as well, for if reasoning were limited to crisp concepts, we would have little use for it outside pure mathematics.

In what follows, we shall discuss some of the ways in which one can reason with vague information, culminating in the puzzle known as *sorites* or the *paradox of the heap*, which dominates many academic discussions of vagueness. In this chapter, I will present the history of the paradox and the different shapes that it may take. A discussion of how best to explain the paradox is postponed till Chapters 8 and 9.

Reasoning with vague concepts

There is a sense in which vague concepts are parasitic on comparative ones: the concept 'tall', for example, would not make sense if we could never tell whether one person is 'taller' than another. Once a dimension of comparison has been established and a measurement method is in place, comparisons are crisp: either John is taller than Bill or he is not.

Understood in this way, the logical properties of comparisons are not difficult to capture, though a few things are worth noting. First, people have a tendency to arrange things in straight lines, as it were, rather than in circles, as when chairs are arranged around a round table. Distances, for example, are arranged in the same way as the real numbers. Weight, temperature, the pitch and amplitude of sound, and so on, are all ordered in the same linear fashion. In all these cases, we appear to think about matters in terms of there being *more* or *less* of something: more distance, more weight, and so on. This is even true for time: despite the circular form of our clocks, we do not measure quantities of time cyclically. *Circular* orderings—such as the compass points that we use to measure wind directions—are interesting, but they do not present us with deep puzzles other than the ones that affect *linear* orderings as well. Not much will be lost by disregarding them here.

Secondly, suppose we focus on height measurement, assuming that we have settled on some well-defined method: a measurement is performed and the outcome is rounded in some standard way. It is now quite possible that two people are assigned the same height, in which case neither of the two is taller than the other. In this respect 'taller' is a bit like 'greater than' (>) when applied to numbers, but with a difference: in the case of numbers, the only case in which neither $a > b$ nor $b > a$ occurs when a and b are the same number. In mathematical jargon, the relation 'greater than' is *connected*. English comparatives such as 'taller' are not connected, because two *different* people can have the same height. Because of this complication, the logic of comparatives is trickier than comparisons between numbers. As it turns out, however, matters can be neatly summarized in two laws, which are sometimes known as *asymmetry* and *pseudo-connectedness* respectively. I present them in English rather than in symbols, writing 'more X than' as a placeholder for any comparison that one might care to make (more powerful, more short, and so on).

1. Suppose a is more X than b.
   Then b cannot be more X than a.
2. Suppose a is more X than b.
   Then any c must be less X than a or more X than b (or both).

All that matters about comparatives can be derived from this by logical deduction. It follows, for example, that in the case discussed above, where neither of a and b is more X than the other, a and b must compare in exactly the same way to any other object c (in the sense that any c that is more X than a must also be more X than b). We shall see later that this is no longer the case if we turn to *vague comparatives*, as when we say 'John is *significantly* taller than Bill'.[2]

Crisp comparatives, then, are not very problematic. So how about the vague concepts that piggyback on them? In what follows, we focus on

128

gradable adjectives that function in isolation, without a noun. In other words, we shall be discussing such expressions as '*x* is so-and-so' rather than, for example, '*x* is a so-and-so man'. This will keep our lives simple, and it will not distort matters, because a very similar story can be told about more complex expressions that contain nouns as well as adjectives.

When I say that some gradable adjective applies, by a first rough approximation, I tell you that there is more (or less) of the relevant quality than some threshold. By saying that the weather is hot, I say that there's more heat *than normal*, or more *than I like*, for example. By saying that John is short, I say that he is less tall than some threshold *t* (which may be my own height, people's average height, etc.). We identify a quality (heat, height, etc.) and say there's more than *t* of it, or less than *t*. What exactly *t* is is often left a bit unclear: we assume that the context will tell us what *t* is.

There may be exceptions to this rule. When I say that the temperature is normal I am not saying that the temperature exceeds some given temperature, nor am I saying that it falls short of some temperature. Rather, I'm saying that the temperature is *close* to some given tempera-ture. This suggests that we can also use adjectives to say that there is neither much more than *t* nor much less than *t* of the relevant quality. Be this as it may, we shall focus on the most frequent adjectives, which either say that there is more or that there is less of something than some contextually given threshold. There appear to be three 'laws' that govern the use of a degree adjective: admissibility, non-transitivity, and toler-ance. We discuss them one by one.

Law of admissibility

This first rule is simple and uncontroversial. It is best explained by focusing on a concrete predicate: 'short', for example, as related to the height of a person. Admissibility says that, if a given object *x* is short, and another object *y* is even shorter than *x*, then *y* must also count as short. Conversely, if an object *u* is not short, while another object *z* is taller than

u, then it follows that z is not short either. This law makes explicit that such concepts as 'short' and 'not short' are at opposite extremes of the height scale.

Law of non-transitivity

This law is best thought of in the negative. Suppose x is smaller than y. This relation is *transitive*, as logicians say,[3] because if x is smaller than y, and y is smaller than z, then x must always be smaller than z. There is a smooth transition here from the first pair (involving x and y) and the second pair (involving y and z) to the third (x and z). Many important relations are transitive: 'older than', for example, is transitive, and so is 'ancestor of'. The relation 'father of' is not transitive, however, nor are relations such as 'know' or 'love', as one may easily verify: the fact that Karenin loved Anna and Anna loved Vronsky does not mean that Karenin loved Vronsky.

Now let us consider equality and ask whether it is transitive. John's children and Leo's children may be equal in number. The relation involved is transitive: if it is also true that Leo and Tim have the same number of children, then so do John and Tim. But here is the rub: vagueness destroys transitivity. Consider, for example, the relation of *approximate* equality. This relation is not transitive. For if x is approximately equal to y, and y to z, does it follow that x is approximately equal to z? Of course it doesn't. It will be useful to see why exactly this is.

What do we mean by 'approximately equal'? Dealing with distances, we might take this to mean 'within 1 cm of each other'. But if x is within 1 cm of y, and y is within 1 cm of z, then x and y may be related to each other in various ways. It is possible that the difference between x and z is still within the 1 cm margin, it might even be nil. But equally, the difference between them might exceed the 1 cm margin, up to as much as 2 cm (see Fig. 7).

So, unlike strict equality, approximate equality is not transitive. The same is true for the relation of indistinguishability, for which we shall use the symbol $\sim$. I may not perceive a difference between x and y, or

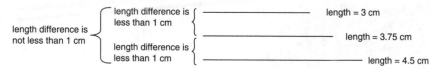

FIG. 7 Two lengths approximately equal to a third one need not be approximately equal to each other

between y and z, yet I may well perceive a difference between x and z. The sum of the two differences may well be big enough to enable me to tell the two apart.[4]

Some[5] have questioned whether there is genuine non-transitivity here, because we are talking about three different measurement events. Consequently it is possible, for example, that the y in the first measurement (when it is compared with x) gets a different reading from the y in the second one (when it is compared with z). So, when we summarize the story by saying that $x \sim y$, $y \sim z$, but $x \not\sim z$, we forget that the two occurrences of y are different things, and similarly for x and z. The idea to distinguish between different measuring events is important, and we shall return to it in Chapter 9. But the argument against transitivity can take this complication in its stride. One way to see that transitivity does break down is by aggregating together *all* measurements of a given thing: we choose x and y that are close enough together that they come out the same on *all* measurements by a given perceiver, and the same for y and z. Let us write $x \sim\sim y$ instead of 'x yields the same reading as y on all measurements'. One can now show that $x \sim\sim y$ and $y \sim\sim z$ do not imply $x \sim\sim z$. Stronger even, one can choose x, y, z so that $x \sim\sim y$ and $y \sim\sim z$, yet x differs from z on all readings.

To see this, let's do a small experiment using a weighing balance. A balance is an excellent model of indistinguishability, because it tells you whether two things are equally heavy: the bar from which its two weighing pans hang is horizontal if both pans contain the same weight (see Fig. 8).

Every balance has limited sensitivity. The balance I am using—a nice antique one made by John Drew of Glasgow—was very precise for its

FIG. 8 The balance used in the experiment

time, but it is not sensitive enough to register the removal of just one grain of rice from one of its pans: the friction between the various metal parts of the balance would overwhelm the weight difference. Suppose we agree that one thing counts as heavier than another if the pan holding the former sinks entirely to the bottom, so that it comes to rest stably on the wooden base of the instrument. (Any other criterion will do, as long as it does not allow misunderstandings or borderline cases.) Using this criterion, it turns out that I need about 35–40 grains of rice to make the balance register the difference between the two pans. The margin of error of 5 is because the balance will behave *slightly* differently every time you use it, because of moisture, the varying positions of rust particles, and so on. All the same, it turns out that 25 grains are never enough to tip the balance, while 50 grains will do the trick every time. So if x is an amount of 10 grains of rice, y is 35 and z is 60, then the balance *never* registers the difference between x and y, nor the difference between y and z, but it will *always* register the difference between x and z, which exceeds 40: the pan containing z sinks to the bottom while the other one is lifted into the air. Much better instruments are available these days, of course, but the same pattern can always be observed. The same is true for our eyesight, our hearing, and our other senses, all of which have limited sensitivities.

In the next chapter, we shall subject the notion of indistinguishability to closer scrutiny, asking how the context in which objects are compared can make a difference. Although this will enable us to use different notions of indistinguishability (some of which are transitive), the basic observations of the previous paragraph will remain unaffected.

Law of tolerance

The third law of vagueness is the most controversial of the three. It stems from the philosopher and linguist Hans Kamp, who based it on the notion of an *observation* predicate. An observation predicate is the kind of predicate that directly reflects an observation or impression. Suppose, for example, you are presented with two paintings that are so incredibly similar that you cannot see any difference between them. Now suppose you find one of the two paintings beautiful, perhaps to the point that you are willing to part with good money for it. Now if 'beautiful' is an observation predicate, then you will have to concede that the other painting is also beautiful (even if you know it is a forgery while the first one was an original). After all, you see no difference. The word 'short' is an observation predicate, since it describes the impression that we have of a person's height. Kamp proposed the following principle of the equivalence of observationally indifferent entities (EOI), which has come to be known as the law of tolerance, or the tolerance principle. Suppose P is an observation predicate, then the following must hold according to this law:

> *Tolerance principle*: suppose the objects a and b are observationally indistinguishable in the respects relevant to P; then either a and b both satisfy P or else neither of them does. (Kamp 1981)

In other words: as far as *observation predicates* are concerned, if you cannot observe a difference then there *is* no difference.

To see what Kamp was driving at, suppose you showed me a person and asked me, 'Is this person short?', to which I say yes. Now you show me a second person, whom you have carefully selected to be slightly taller than the first, but only by so little that I am unlikely to see the difference. You ask me two questions about this second person, who is standing next to the first: 'Can you see the difference from the first person?' Suppose I say no, because I am unsure which of the two is taller. Now you go for the 64,000-dollar question: 'Is this second person short?' Surely, it would be strange, under these assumptions, if I said no. If I view the first person as short then the same must be true for the second person.

The tolerance principle does not hold for all properties. It does not hold for words such as 'obese', for example, as sanitized by the notion of body mass index. We have seen that 'overweight', for example, is often defined as a BMI between 25 and 29.9. This makes someone whose BMI is 25 overweight, but someone whose BMI is 24.999 not. Similarly, if we use the standard relative definition of poverty, someone who earns 60 per cent of the median income does not count as poor, while someone who earns one penny less does. These formalized notions of overweight and poverty are *not* observation predicates.

The sorites paradox

In ordinary life, the word 'paradox' can mean pretty much anything we find puzzling: if someone is in a melancholy mood on his wedding day then this is paradoxical. House prices go up while consumer prices go down? Paradoxical! But mathematicians—and most other scientists in their wake—use the word in a stricter sense. To them, a paradox is a situation in which a statement is proven, and so is the negation of that same statement. Something is made out to be true and false at the same time, in other words. It is in this most worrying kind of paradox that reasoning about vagueness entangles us.

134

When paradoxes occur in a mathematical system, the situation is grave indeed. Mathematicians, at their most meticulous, work with precisely formulated reasoning systems, which describe how certain assumptions (called *premisses*) lead to precisely specified conclusions. Many areas of mathematics have been put on a formal footing along these lines, allowing mathematicians to prove things with great rigour. From time to time, a system of this kind is shown to lead to contradictory conclusions: a pair of statements p and $\neg p$ (i.e. not p), both of which are proven from the same axioms. A paradox, in this precise sense, is not just puzzling: it is catastrophic. Reasoning systems are usually constructed in such a way that absolutely *anything* will follow from a contradiction. So once a contradiction is proven, you can prove *any* statement that can be formulated in the formal language employed by the system, making the whole reasoning system useless as a result. The system becomes like an oracle that responds 'yes' to any question that you throw at it.[6]

In this book, we are concerned with paradoxes arising in ordinary language. The ancient Greeks produced many paradoxes of this kind. In the best ones, all the premisses and inferences appear immaculately sensible. A famous example is Zeno's proof that speedy Achilles cannot overtake the slow tortoise: when he arrives where the tortoise starts, the tortoise has progressed to some later point; when Achilles arrives there, the tortoise has progressed to some still later point; and so on and so forth. Paradoxes like this can teach us important lessons. The paradox of Achilles and the tortoise, for example, appears to have played a role in teaching scientists to discard incorrect models of space and time. Lessons can be learned from sorites as well. Let us start by reiterating one of its best-known versions, in more detail than before.

0. 100 stones can form a heap.
1. 0 stones cannot form a heap.
2. For every n, if n stones cannot form a heap then $n + 1$ stones cannot form a heap.
3. Therefore, 100 stones cannot form a heap.

Since stones come in different shapes and sizes, which can be stacked together in different ways, this version of the paradox is a bit messier than others. But let us pay our respects to history and stick with the stoneheaps, which make their point handsomely enough. The argument is a paradox because (3) contradicts (1). It is important to note that (2) could be replaced by 100 different sentences, each of which is of the form 'if ... then ...':

> If 0 stones cannot form a heap then 1 stone cannot form a heap.
> If 1 stone cannot form a heap then 2 stones cannot form a heap ...
> If 99 stones cannot form a heap then 100 stones cannot form a heap.

Although the original rendering is more succinct, the latter one is constructed from simpler material. To understand it, you do not have to understand difficult concepts such as 'every': all you need is sentences of the form 'if ... then ...', which are called *conditionals*, or 'implications'. We shall focus on this elaborate form of the paradox.[7] We will often have occasion to refer to the conditional statements that lie at the core of this version of the paradox, calling them the *sorites conditionals*.

Little is known about the history of the paradox, but from ancient times onwards, Eubulides of Milete has been thought to be its inventor.[8] Eubulides lived in Aristotle's times, and the two knew each other's work. Some sources suggest that their exchanges were rather acrimonious: a fourth-century bishop named Eusebius even claimed to have read a book by Eubulides that charged Aristotle with spying. Unfortunately, the book seems to have disappeared off the face of the Earth. What Aristotle thought of the paradox is unclear.

Why was the paradox called 'sorites' anyway? It may well have been because the paradox was first applied to the notion of a stoneheap: *soros*, in ancient Greek, means 'heap' or 'multitude', as is generally thought. There is, however, an intriguing alternative explanation, which hinges on the fact that Eubulides was familiar with Aristotle's

136

calculus of syllogisms. A syllogism is any of a number of patterns of inference revolving around such words as 'all' and 'some'. For example, in a slightly modernized notation:

If all A are B, and all B are C, then all A are C.

(For example, if all Sicilians are human, and all humans are mortal, then all Sicilians are mortal.) Now the Greek verb *soreuo* means 'to stack together' or 'to string together'. In later centuries, 'Aristotelian sorites' was a name for a slightly more complex version of the same argument in which different syllogisms are strung together.[9] For example,

If x is an A, all A are B, all B are C, and all C are D, then x is a D.

It is unclear whether stacked arguments were called 'sorites' when Eubulides named his paradox, but if they were then he may have chosen the name because the paradox hinges on the stringing together of a large number of apparently harmless reasoning steps. In the Aristotelian sorites, there is no vagueness; each step is completely crisp. It does not matter how many syllogisms are strung together: the result is always a valid argument. Eubulides' own argument is very different, of course, because it becomes less plausible with each new addition:

100 stones can form a heap. [*Of course they can.*]
0 stones cannot form a heap. [*Of course not.*]
If 0 stones cannot form a heap then 1 stone cannot form a heap; therefore 1 stone cannot form a heap. [*Yes indeed.*]
If 1 stone cannot form a heap then 2 stones cannot form a heap; therefore 2 stones cannot form a heap. [*Indeed.*]
If 2 stones cannot form a heap then 3 stones cannot form a heap; therefore 3 stones cannot form a heap. [*It depends...*]
...
If 20 stones cannot form a heap then 21 stones cannot form a heap; therefore 21 stones cannot form a heap. [*That's nonsense!*]

Eubulides might have reasoned that his problematic argument involved a stringing together of reasoning steps just like an Aristotelian sorites, but in such a way that the stringing together makes it gradually less valid. Steven Pinker once likened the process to that of repeatedly copying an analogue audiotape: the copying process is faithful enough, yet a tenth-generation bootleg tape sounds nothing like the original. In Pinker's aptly chosen words: the slop accumulates.[10]

Vagueness as ignorance

There is a school of thought[11] that denies the plausibility of the paradox altogether, roughly along the following lines. (See also the dialogue intermezzo on this topic.) Suppose you confronted a viewer with collections of same-sized stones, all stacked together in heap-like fashion, but with different numbers of stones. If the viewer is always forced to make a choice between calling something a stoneheap or a non-stoneheap then, surely, each collection would either be called a stoneheap or not. There are no two ways about it (or so the argument goes): after the experiment, we are able to draw a precise line between the things our viewer considers a stoneheap and the things she does not. Before doing the experiment, we may not know where the line will be drawn, but it will be drawn *somewhere*. One of the crucial premisses of sorites is incorrect, since there is a value n such that n stones do not make a stoneheap but $n + 1$ stones do.

This is known as an *epistemicist* response to the paradox, because it sees the trouble with a vague expression as arising from our lack of knowledge (Greek: *episteme*) of a crisp boundary that does nevertheless exist: epistemicism sees our ignorance as the root of the problem. Responses of this kind are sometimes summarized with the slogan 'vagueness as ignorance'. In its simplest form, epistemicism treats the case of the stoneheap a bit as if it involved, say, the money in your bank account. Suppose, coming back from a holiday, you do not know how much money you have in the bank. You start withdrawing money very cautiously, one penny at a time. No money gets added to your account, so your money will decrease steadily, and overdrafts are not allowed. Sooner or later, one of your withdrawals will be rejected, but you do not know when. The epistemic approach says that this is a good metaphor for the story of the stoneheap, where you do not know where to draw the line between stoneheaps and non-stoneheaps. The line exists, and it exists at some precise point, we just do not know where.

Theories of this kind do have the virtue of simplicity since, by denying the crucial premiss or premisses, they allow their proponents to treat vague predicates more or less as if they were crisp. As a result, they do not have to consider fancy new logical systems: classical logic—an approach to logic whose outlines will be sketched in the next chapter—will do. In some authors this position is accompanied by some fine rhetorical flourishes. Roy Sorensen, for example, wrote in this connection, 'The history of deviant logics is without a single success.' While such a position may have looked defensible in the 1950s, it is not any longer. I suspect that, after the successes of some highly 'deviant' logical systems, it has become difficult to find a logician who agrees with Sorensen.[12] Counter-examples include, for example, any of a large number of *non-monotonic* logics, which deviate from classical logic quite drastically by relinquishing the basic rule which says that a conclusion, once derived, can never be retracted on the basis of new information (i.e. additional premisses).

The prevailing attitude among logicians, these days, is to see logic as a tool for a job, and to acknowledge that different jobs require different tools. Sorensen's plea for always choosing the simplest tool that will do the job is well taken. It is starting to appear, however, that the sorites paradox may require some deviations from classical logic. Sticking with a tool that we know and love might be a luxury that we cannot afford.

If the epistemicist approach has a certain appeal in the case of the stoneheap, I find it less convincing in other situations. Consider:

Premiss 1: A person of 150 cm is short.

Premiss 2: (for all natural numbers n,) if a person of n cm is short then a person of n cm + 0.0000001 cm is short.

Conclusion: Any person of at least 150 cm is short.

Imagine an epistemicist saying: 'Premiss 2 is false. There does exist a number n such that if you're n cm then you're short but if you're 0.0000001 cm taller then you're not! Which value of n is that? No one knows.' One difficulty with this view—which its adherents have attempted to address—is that it does not explain why we find Premiss 2 appealing. Another is that it implies that there is an essential aspect of the word 'short' that speakers of English do not know, and which they cannot know, if only because their eyes have limited resolution. This seems acceptable in connection with such words as 'water' and perhaps even 'obese'. In the case of water, for example, the nature of the beast is something that needs to be discovered through chemical research; for such words as 'short', however, it is difficult to see how research can matter.

Epistemicists can counter that, in cases such as 'short', it is the behaviour of the language community as a whole that determines their meaning. For surely, a child learns to use words in the same way as the people around it. This in itself is a reasonable position, but it is not without problems, for example because different people use words in different ways. Consider weather forecasting, for example. Ehud Reiter and others at the University of Aberdeen noticed that weather reporters used words such as 'evening' in very different ways.[13] When confronting reporters with these differences, all manner of relevant factors came to light. Some forecasters, for example, felt that evenings start later in summer than in winter, while others thought the season was irrelevant. Similarly, some thought that dinner time had something to do with it, whereas others felt such trivial matters to be irrelevant. At least in the case of 'evening', the idea that speakers converge on the same interpretation does not have a great deal of empirical support. So if epistemicism has a point, it will somehow have to make its peace with the facts of variations between speakers.

Clearly, sorites arguments are not just a matter of language: they hinge on the way in which our senses work. To obtain a better understanding of some of the relevant mechanisms, let us examine the case of colour.

Similarity is in the eye of the beholder: The case of colour

Any definition of 'red' which professes to be precise is pretentious and fraudulent.

RUSSELL (1948)

Rohit Parikh reports on a small experiment in which thirteen subjects were shown a part of the Munsell colour chart, containing squares of different colours. Participants were asked to write down the number of blue squares and the number of red squares that they saw. There was no consensus at all. Of the thirteen respondents, not even two reported the same number of blue squares, for example, varying in their responses between 0 and 27. (Similar results are reported elsewhere by Moxey and Sanford (2000) concerning words such as 'many'.) Parikh goes on to show that not all is necessarily lost if we do not always understand a word in the same way as the people with whom we talk. He starts from the observation that different people understand colour terms somewhat differently, but simplifies matters by assuming that everyone's understanding must always be crisp.

Ann and Bob, who live together, work at a college where Ann teaches maths and Bob history. One day around noon, Ann is in her office while Bob is working from home. Ann calls Bob asking him to bring her book on topology, an area of mathematics, when he comes in after lunch. Ann has only one topology book, and Bob will know the book when he sees it. Bob asks, 'What does it look like?' and Ann says 'It is blue.' 'Blue' is not precisely defined, so when Bob starts checking all the books that he calls blue, he

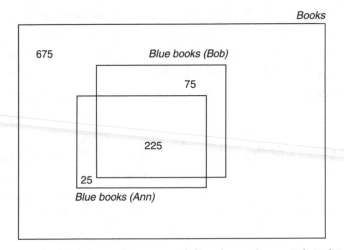

FIG. 9 Blue books, according to Ann (left) and according to Bob (right)

may be lucky and find the topology book among them. But if he is unlucky, the book is one of those Ann calls blue but he does not. If this happens, he may have to check all Ann's books. In Parikh's example, Ann owns 1,000 books. She calls 250 of them blue, and Bob 300, 225 of which are also among the ones Ann calls blue (see Fig. 9).

Parikh now shows that, on the balance of probabilities, Ann's imperfect instruction will nevertheless save Bob time. Let's see why. If she had not giving him any help, he would be looking at as many as 1,000 books. On average, he would have had to check about half of them before hitting on his target. Statisticians would say Bob's *expectation* is 500. Given Ann's instruction, this expectation goes down to 200, as Parikh shows, because it is very likely that the book will be among the 300 books Bob calls blue; he will only have to inspect the other 700 in those 10 per cent of cases where it is not. It would have been better for him if Bob had understood colour terms in exactly the same way as Ann, of course, but her instruction is useful nevertheless.[14]

There is not always such a happy ending (as those of us know whose private lives resemble those of Ann and Bob). Mismatches in the way we understand words can be serious. If Ann and Bob fail to agree about

142

the colour of any book, for example, then Ann's description of the book as blue is actually detrimental, because it forces Bob to first inspect *all* the books he himself calls blue, without finding it, after which he can try his luck with the other ones. (If, as in the original story, Bob calls 300 books blue then this means he should expect to check as many as 650 books.) To what extent people agree on the meaning of words when they talk about topics in public life, such as failing teachers, poisonous substances, or obesity for example, is anybody's guess.

It will be useful to look at the mechanisms behind colour perception in more detail.[15] The colour of an object is determined by what percentage of light is refracted from it at each wavelength. But our eyes perceive colour quite economically, by paying attention to just three overlapping ranges of wavelengths, known as the red, green, and blue range. (The meaning of these terms does not coincide with normal usage, which is why we shall put them in quotation marks below.) Our eyes operate by *summarizing* what they find in each of these three ranges. Suppose you are looking at a particular patch of colour. Far from passing on all the individual values, your eyes take the average of all the values that it measures within each range. All that is passed on to the brain is three numbers: the percentage of 'red' light refracted, the percentage of 'green' light refracted, and the percentage of 'blue' light refracted. If you are like most of us, these three figures will allow you to separate about 10 million different colours, some of which are grouped together by everyday words such as 'blue', 'pink', 'red', and so on. Needless to say, you are unlikely to have a word for each of them: even standard works on colour terminology do not contain more than a few thousand. To complicate matters further, different languages name different parts of colour space: Russian, for example, has no name for what the English-speaking nations call 'blue', essentially dividing it into light blue and dark blue.

So far so good. But summarization means loss of information: it causes some colours that are physically quite different to appear exactly the same. We are not talking about objects that have *subtly* different refractions: the fact that such objects can be indistinguishable is obvious. But consider the 'blue' range, for example. An object *a* may have low refraction percentages in the lowest part of that range, increasing gradually throughout the range; object *b* may behave in the opposite way, starting with high refraction percentages in the lower parts of the blue range and going downhill from there. The difference between *a* and *b* at the low end of the range can now be cancelled out by the difference at the high end, so both objects may end up with the same *average* blue refraction, causing their 'blueness' to be represented by exactly the same number. The objects *a* and *b* are known as *metamers*: colours that are indistinguishable even though they are physically very different. Metamers make indistinguishability particularly interesting in relation to colours, but they do not change the logic of the concept. Indistinguishability remains non-transitive, for example: of three shades of green *a*, *b*, and *c*, it may well be that *a* and *b* are indistinguishable from each other, and so are *b* and *c*, yet *a* and *c* may be different enough to look different. The practical implications are familiar, namely that artificially imposed boundaries between colours could never be applied with absolute precision.

Up to this point, we have been speaking as if one and the same notion of indistinguishability applied to all people. And indeed, it is easy enough to look up, for any two shades of colour, whether they are counted as distinguishable or not, by consulting the *standard observer* model of the International Commission of Illumination. But the existence of such a standard is deceptive, because it is only a kind of average between normally sighted observers: 'A well-chosen standard of this sort although not accurately describing all or even most normal perceivers will not differ too much from the perceptions of any particular person' (Hilbert 1987). As Hilbert explains, the reasons why different people can distinguish different pairs of colours are well understood: they are caused by factors such as the density of pigment layers on the lens and over the central area of the retina, and by the different sensitivities of the photoreceptors in the eye. Such differences can exist between people who are equally good at telling colours apart. I do not know whether the same sorts of difference are found in simpler domains, but anecdotal evidence suggests that this is possible. My mother, for example, is quite good at telling people's heights apart, as long as they are in the 185–200 cm range. Outside that range she's a little hazy.

The story of colour perception suggests that the differences between the ways in which people use a given colour term are unlikely to be ironed out as a result of social interaction: our eyes are too different. What is visible to you may not be visible to me. It would be unreasonable to expect a blind speaker to learn the same colour concepts as a seeing person. It would be almost equally unreasonable to expect the same between normally sighted people. Complete convergence between speakers must be ruled out.

DIALOGUE INTERMEZZO
ON VAGUENESS AS IGNORANCE

Following an exam at the end of their beginners' course on Spanish, Roy and Hans exchange opinions about the exam. Their joint interest in the philosophy of language makes itself felt, so before they know it, they find themselves discussing the virtues of epistemicism, a position also known as 'vagueness as ignorance'.[16]

HANS: *How did you translate 'alto'?*

ROY: *'Tall'. That's what it means, I think, when applied to a person.*

HANS: *Funny, isn't it? At work, we're meticulous about every word and symbol. But when we're learning a language, we're content to be told that 'alto' means 'tall'. We're not asking: 'How tall does a Spaniard need to be before they call him "alto"?'*

ROY: *Which is fair enough, isn't it? You board a plane to Madrid, you look around, you get a pretty good idea who is 'alto'. A few examples of how Spaniards use the word will do the rest. You won't be able to say exactly where the boundary lies, but you won't be far off either. You go to Mexico City, same story, except you'll be approximating a different threshold, probably slightly lower.*

145

HANS: *Your word 'approximating' worries me. You talk as if there's something definite to be approximated there: a precise boundary separating the tall ones from the others. You're talking a bit as if heights were all as neatly separate from each other as the species of animals in a zoo. And even in the case of animal species, such separateness cannot be taken for granted...*

ROY: *Yes. A potentially different boundary at each place and time, but yes, a precise boundary.*

HANS: *Surely you do not mean really precise? You are not saying that a person just one billionth of a millimetre above the boundary is tall, whereas someone who is below the boundary by that same puny distance must necessarily count as not tall—do you?*

ROY: *That is exactly what I do mean.*

HANS: *But that's a boundary that no human being could ever apply with any confidence.*

ROY: *That's right. Too bad!*

HANS: *How on earth do you know that such a boundary exists?*

ROY: *Well, I find it difficult to conceive of any other arrangement. Surely some pattern must underlie speakers' and hearers' use of the word.*

HANS: *But what kind of pattern? One thing I'd worry about is inconsistencies in usage. Take words like 'afternoon' and 'evening'. People differ over whether 6 o'clock is in the evening.*

ROY: *Sure, inconsistencies between speakers can arise. Five hundred years ago, words like 'fat' and 'poor' were used inconsistently. But science shows us the best way to use these words.*

HANS: *And you believe the same will happen to words like 'tall'? I doubt it. But suppose you're right. That would constitute a change in the meaning of these words. If inconsistencies existed in people's use of language in the past you can't paper them over by pointing out that current speakers use these terms consistently.*

ROY: *Maybe we should view the newfangled usage as correct, and earlier usage as a rough approximation.*

HANS: I'm sure the same is true for our cave-dwelling ancestors: they were all aspiring scientists, weren't they? If Fred Flintstone submitted that Wilma took a bit long to cook his dinner then his usage of the word 'long' was...

ROY: Maybe I was getting a little too enthusiastic. But the epistemicist's point is really this: people try to make themselves understood. Therefore, they must aim to converge on the same meaning.

HANS: Aim...sounds reasonable.

ROY: So this might make the existence of one 'correct' meaning a useful fiction.

HANS: A useful fiction, yes perhaps. But why this correct meaning would have to be a crisp dichotomy is not clear to me...By the way, something else bothers me about epistemicism, quite apart from the trouble with inconsistent usage.

ROY: What's that?

HANS: Lack of usage. Data sparseness if you like. Suppose I coined a term right now, let's say 'flibberiness', which did not exist before. Suppose, from now on, I use this term to denote the sensation in my mouth caused by eating rhubarb.

ROY: I know the feeling.

HANS: Good. It can be stronger or weaker of course. Epistemicism posits that, right now, by coining the word, I have automatically defined the exact boundary at which the feeling is strong enough to say that I'm feeling flibbery (somewhat flibbery, utterly flibbery, and so on). It's a mystery to me how this boundary has come about in an instant.

ROY: Maybe epistemicism applies only to well-established concepts.

HANS: That's quite a concession to make! New usage is quite common. Consider a word like 'quickly'. Wilma has invented a new cooking recipe, dinosaur ears with olives. Today she's cooked it rather quickly. New usage, new boundaries. Plus, if there exist entirely new concepts then presumably there exist somewhat new concepts, which must suffer from the same problem to a smaller extent.

ROY: Perhaps you're right.

HANS: *Really, the more I hear about epistemicism the more sceptical I become. The meaning of an expression is created by its use, right? If that's true then there's not a lot we can take for granted about its meaning without inspecting its use. If the expression is used a lot then there are bound to be inconsistencies. If it isn't then there will be gaps. In both cases, goodbye boundary!*

ROY: *Perhaps I was putting the cart before the horse when I suggested that people approximate a threshold. A model approximates reality, not the other way round. Take an epidemiologist trying to correlate two factors: amount of exercise, for example, and probability of heart disease. The more exercise the fewer heart attacks, let's say. To get a good view of his data, our epidemiologist draws a diagram: a scatter-plot as they say. The plot is a bit messy; still he might manage to correlate the two factors by means of some function. Drawing it into his graph, he might see a straight line, for example, which connects many of the dots.*

HANS: *I know what you mean. I've heard of linear regression.*

ROY: *Let him use what technique he wants. My point is: no one in his right mind expects to get the correlation exactly right. There are bound to be some dots above and below the line, and there will be a lack of data in some parts of the graph as well. Still, the line might be an excellent approximation.*

HANS: *And your point would be?*

ROY: *...That a linguist looking at our use of the word 'tall' might be able to posit a boundary between tall people and others. Once again, the boundary is not exactly correct. Some speakers would be better modelled by a somewhat lower boundary and others by a higher one. It's only an approximation of people's usage.*

HANS: *All right: the boundary approximates usage...*

ROY: *That's it. Not the other way round.*

HANS: *Or maybe in both directions? Because you do have a point, that speakers try to converge with each other... Anyway, I have little doubt that a simple*

148

> *model of the meaning of 'tall', which treats it as a simple dichotomy, can go a long way.*
>
> ROY: *You would have to use a different boundary when measuring children or adults, of course, even men or women, Spaniards or Mexicans, and so on, but I don't need to tell you that.*
>
> HANS: *Of course. And that's fine with me. The point is that simplifications can often work pretty well. I've got no quarrel with them. As long as no one is claiming they're absolutely correct!*

The idea that theories can be seen as approximations of the facts is an important one. The various theories in the next two chapters can be seen as attempts to construct increasingly accurate approximations of the way in which vague expressions are used. Before we go there, however, it will be important to know in what situations the sorites paradox comes up. We will discuss this issue by exploring the different meanings of the word 'continuity', which is often linked with vagueness. 'Continuity' is used in a number of different senses, so let us make an attempt to unravel this knot.

Continuity and vagueness

One of the earliest accounts of continuity is ascribed to the seventeenth-century philosopher and mathematician Leibniz. The relevant principle of continuity asserts that everything in nature happens by degrees: 'Nature makes no leaps.' The idea seems reasonable enough in combination with a wide range of naturally occurring phenomena. It would be unexpected, for example, if the Earth contained mountains of all sizes below 8,000 metres, except between 2,000 and 4,000 metres. Similarly, if there are days when the maximum temperature is 20 °C and days where the maximum is 40 °C then one would expect to see days with a maximum of 30 °C as well. Or take change over time: a plant or a fish or a baby does not grow to its full size in an instant, but via

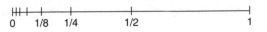

FIG. 10 Density: between any two fractions lies another fraction

numerous intermediate stages, each of which differs only marginally from its predecessor.

This early version of continuity, however, is imprecise, and imprecise ideas are not easily tested. Consider the growth of a plant, for example: if plants grow by cell division, and if cell division is more or less instantaneous, does this constitute a violation of the principle? Or, does the role of quanta in modern physics falsify the principle, because quantum physics models energy as coming in separate packet? Such questions could be answered only if the principle was formulated more precisely.

A mathematically precise version of continuity was proposed around 1900. Its natural *habitat* is the mathematical world of 'structures': sets of objects with relations defined between them. *Numbers* come to mind, ordered by relations such as 'smaller than'. Mathematical continuity can be seen as a refinement of the technical notion of *density*, which is easier to define: the familiar < relation as defined on the set of all fractions, for example, is dense because in between any two fractions there always exists a third one. For example, one can always choose the average of the two: in between 0 and 1 lies ½; in between 0 and ½ lies ¼; in between 0 and ¼ lies ⅛, and so on (see Fig. 10).

It might be thought that things do not get any tighter than this: if two structures are both dense, how could one of them be even more tightly packed than the other? Enter the concept of mathematical continuity: a mathematically continuous structure is 'denser than dense'.[17] Essentially, these structures are ordered in the same way as the real numbers, which contain such numbers as $\sqrt{2}$ and π, in addition to all the fractions. Continuous structures cannot fail to be dense; in some sense they are as tightly packed as can be conceived. Perhaps surprisingly, mathematical continuity is largely irrelevant to the study of vagueness, because sorites

150

arguments can arise in structures that are not even dense, let alone continuous. Why is this?

What is crucial for sorites, as we have seen, is the existence of elements which resemble each other closely enough that they are difficult to distinguish. If a structure is continuous then it contains elements at arbitrarily small distances from each other. But pairs of highly similar elements can easily exist in a structure that is not even dense. Consider our sorites argument involving a person's height, for example. The structure involved consists of a large set of people, who can be arranged in a linear sequence so that each of them is shorter than the next one. This structure is not dense, because there is always a next person in the sequence. Yet crucially, the difference of 0.0000001 cm between subsequent people was not observable, triggering the tolerance principle. This suggests yet another notion of continuity, known as *psychological continuity*. A structure is psychologically continuous if it involves a non-transitive notion of indistinguishability (i.e. containing three elements a, b, c, such that $a \sim b$ and $b \sim c$, but *not* $a \sim c$). It is this last descendant of Leibniz's principle that is important in connection with vagueness.

In the following chapter, we focus our investigation on those trickiest versions of sorites which exploit domains where psychological continuity plays a role. We do not have to ask whether the world is mathematically continuous or dense. We won't have to immerse ourselves in any hairy physics either, which is just as well.

With these clarifications out of the way, it is time to move on and to ask, more systematically than before, how to treat the problem of sorites.

Things to remember

◆ Gradable concepts (such as size or height) give rise to related notions (such as 'large', 'tall', 'short') that are vague.

◆ Even with vague concepts, rigorously precise reasoning is sometimes possible, but reasoning can also become problematic in some cases. An example of such an imprecise piece of reasoning is the ancient sorites paradox.

◆ Three of the main principles governing reasoning with vague predicates are the law of admissibility, the law of non-transitivity, and the law of tolerance. *Admissibility* is a simple but important housekeeping rule which forces us to respect the rank ordering between objects (e.g. if Oxford counts as large then Birmingham must also count as large, because it's larger than Oxford). *Indistinguishability* is non-transitive because small differences can add up to very significant ones. *Tolerance* says that perception predicates should never separate objects that are perceptually indistinguishable.

◆ The most vexing versions of the sorites paradox rely on the law of tolerance. The resulting paradoxes can even arise in finite domains, which are not continuous (or even dense) in the mathematical sense.

◆ The epistemic explanation of the sorites paradox does not sit easily with experimental results on the perceptual and linguistic differences between people. Moreover, it fails to explain why the crucial second premiss of the paradox has any plausibility. Finally, it is difficult to see how epistemicism could apply to vague terms that have newly been coined, such as our word 'flibbery'.

◆ The study of *colour perception* demonstrates that similarity between colours really is in the eye of the beholder: colours that are indistinguishable to a perceiver can be physically very different, and need not be indistinguishable to another perceiver. Rohit Parikh's case study demonstrates how a mismatch between the way in which a speaker and hearer understand colour terms does not necessarily make these words useless. Perhaps more importantly, it suggests a utility-based perspective that will stand us in good stead in the third part of this book, when we shall ask when it is a good idea to be vague.

◆ The sorites paradox suggests that the normal apparatus of classical logic might be ill suited for reasoning about the gradable notions on which empirical science is based. It is of great importance to investigate how the paradox may be explained and avoided.

8

Parrying a Paradox

Even a child can tell that the conclusion of a sorites-style argument is wrong: there do exist stoneheaps, for example, and there do exist tall as well as short people. Something, evidently, is wrong with the reasoning that led to the paradox. But what exactly? This chapter is devoted to a variety of relatively conservative answers to that question.[1] But although valuable insights are embodied in each answer, each will also prove to have substantial shortcomings. At the end of the chapter, we shall conclude that more potent medicine is required.

These explorations will lead us into theoretical territory, where some readers may not want to follow. We shall meet them again in Chapter 10.

Logic and paradox

How can we tell whether a logical argument is valid? Arguments are usually expressed in ordinary language: this is true in the pub, over the dinner table, and even in a scientific seminar. But when academics *really* want to see whether an argument holds up to scrutiny, they can abandon ordinary language at least to an extent, replacing it by something more like mathematical formulas. The standard academic approach is to 'translate' the original argument into a system of symbols whose properties are well understood, and which is designed specifically for the purpose of encoding and assessing logical arguments.

Symbolic logic, sometimes known as formal logic or mathematical logic, is a standard tool set for assessing the validy of logical arguments. Most documented responses to the sorites paradox have used symbolic logic in one way or another. Yet, it is worth asking whether the route via logic is a reasonable choice. Perhaps vagueness has evolved in situations that were simple enough that no paradoxical conclusions would ever be drawn: the cavemen who gave us our language may have had better things to do than worry about sorites. Additionally, most of us are quite spectacularly bad at logic, as experimental psychologists never stop pointing out. If this is how things stand, it might be misguided to want to use symbolic logic (of all things!) to clarify the structure of man-made concepts.

Sensible though this sceptical view of logic may sound, it appears to be based on an outdated understanding of what logic is. In the words of the Australian philosopher Dominic Hyde, 'If logic is to have teeth it must be applicable to natural language as it stands.' This has less to do with language than with the new outlook on logic that was alluded to in the previous chapter. Logic is now seen more as an adaptable tool: confronted with a task, you may design a logic—or find one ready-made—to perform that task for you. It is important to realize that this new-found adaptability does not make logical argumentation 'subject-ive': there is nothing postmodern about logic. What it means for logic to be adaptable will become clearer once we have seen a few examples of the ways in which logic has been moulded to deal with vagueness.

Responses to sorites may be classified into ones that stay relatively close to logical orthodoxy on the one hand, and more radically innova-tive approaches on the other. The present chapter focuses on the first category, which is why we will start by setting out some of the main principles of what is known as classical logic. The next chapter will discuss more radical approaches, which deviate further from classical logic. All approaches seek to explain two things: why the sorites argu-ment sounds plausible, and why the argument is nevertheless defective.

A crash course in classical logic

Symbolic logic emerged as a fledgling new area of mathematics in the nineteenth century, as a result of work by George Boole, Gottlob Frege, and others. Their efforts built on much older work by Aristotle and the Stoics, which was finally put on a more solid footing by, among other things, the use of variables, a mathematical mechanism unknown to Aristotle. As a result of various innovations that allowed logicians to say precisely what the well-formed formulas of a logical language are (i.e. the *syntax* of the language), and what it takes for a given formula to be true (i.e. the *semantics* of the language), people began to speak of formal or symbolic logic.

Logic became a formidable branch of mathematics around the 1930s, when Kurt Gödel proved a number of key theorems concerning the power and limits of logical systems. The main logical formalism that had evolved at this stage is still known as classical logic. Let us informally describe some of the features of this formal system. Far from aiming to do justice to classical logic in its full richness, we focus on a few themes that will be important in later parts of this chapter.

First, any statement (also called *proposition*) of classical logic is either true or false: true and false are the only *truth values*, as we say. In honour of George Boole's seminal work, this feature of classical logic is said to make the logic *Boolean*. It is part and parcel of this view that no proposition can be both true and false. In other words, regardless of what proposition p is, both of the following must be true:

At least one of p and not p is true (law of the excluded middle).

p and not p are not both true (law of non-contradiction).

Formulas such as these two are constructed from some extremely simple building blocks: letters, such as p, which locate the place of a proposition, and operators (also called connectives), such as 'not' ($\neg$), 'and' (&), and 'or' ($\vee$), which can be strung together to create more complex formulas. Each

156

operator comes with a precise definition of its meaning. According to this definition, for example, a statement of the form 'p or q' (also written as $p \lor q$; each of p and q can be any proposition, simple or complex) is true if p is true; it is also true if q is true; it is false otherwise:

p	$\lor$	q
true	true	true
true	true	false
false	true	true
false	false	false

This operator is known as *disjunction*, or more elaborately *inclusive disjunction*: inclusive because 'p or q' is true if p and q are both true. Its components, p and q, are called *disjuncts*. In the same way, the components p and q of the *conjunction* 'p and q' are called *conjuncts*. A statement of the form p & q is true if p and q are each true, but false in all three other situations. We can now write the two above-mentioned laws as follows:

$p \lor \neg p$ (law of the excluded middle)
$\neg(p \ \& \ \neg p)$ (law of non-contradiction)

One further operator needs to be mentioned, because it plays an important role in the sorites paradox: it can be written as 'if p then q' (also $p \rightarrow q$), and is variously known as the *implication* or *conditional*. Its first component, p, is known as the *antecedent* of the conditional, the second component, q, as its *consequent*. The meaning of the conditional operator can be defined in terms of the earlier operators:

$p \rightarrow q$ is true if and only if $\neg p \lor q$ is true

This deceptively simple definition causes $p \rightarrow q$ to mean something subtly different from what we usually understand when we say 'if p then q'. Let us see how this works. If p is true while q is false then the formula is clearly false, which is how things should be. Suppose I say 'If the match is shown on the BBC then we stay at home', and the match is indeed shown on the BBC but we do not stay at home, then we are acting in breach of the conditional statement. In other words, the conditional is false. The important bit is to realize that $p \rightarrow q$ is true in *all other* cases. For example, if the match is not shown on the BBC then the conditional counts as true, whether we stay home or not. On its classical definition, the conditional has nothing to do with causality or any other inherent connection between p and q. To reflect this limitation, it is also known as the *material* implication.

p	$\rightarrow$	q
true	true	true
true	false	false
false	true	true
false	true	false

To illustrate with some simple numerical examples, let p be $2 + 2 = 4$ and q be $3 + 3 = 6$; this makes $p \rightarrow q$ true, because q is true. The same holds if p is $2 + 2 = 5$ and q is $3 + 3 = 7$, even though the consequent is false, simply because the antecedent is also false. Despite efforts to design more intuitive formal counterparts to words such as 'if', the material implication is still the standard conditional of logical calculi. To find something more intuitive, you need to use a mathematically more complex calculus than classical logic. For our purposes in this book, this won't be necessary: the material implication will be good enough for us.

What we have seen so far is the part of classical logic that is known as propositional logic, because it focuses on relations between propositions. Things become more complex when *quantifiers* are added to the language, to allow the logician to express sweeping statements about all the objects in a certain domain. For example, 'For all x, if x has the property F then x has the property G' can be written as

$$\forall x(F(x) \rightarrow G(x))$$

Because this extension to propositional logic hinges on separating a *predicate* (i.e. a quality that an object may have or lack) from the object itself, called the *argument*, to which the predicate is applied, it is known as predicate logic. Predicates are true or false of objects: in the domain of numbers, 'even' is a predicate, and so is 'divisible by 4'. In the same Boolean spirit as before, classical logic takes predicates, which are often abbreviated using letters such as F and G, to be either true of a given object or false of it: borderline cases are ruled out. If $F(x)$ stands for 'x is divisible by 4' and $G(x)$ for 'x is even', then the formula above says that all things divisible by 4 are even, which happens to be true. The formula resulting from a swap between F and G would be false, however, because some even numbers are not divisible by 4.

Right at the heart of classical logic, finally, is the idea that one formula can *follow* from one or more others. A formula, say ψ, follows from another formula, say φ, if the truth of φ automatically carries with it the truth of ψ, in other words, if φ cannot be true unless ψ is also true. The formula φ is called the *premiss* of the argument and ψ its *conclusion*. For example, $p \lor q$ follows from p. Sometimes two things follow mutually from each other. For example, $\neg\neg p$ follows from p, and the other way round as well. The idea of one thing following from something else extends naturally to arguments involving two or more premisses: ψ follows from the combination of φ_1 and φ_2 if the collective truth of φ_1 and φ_2 automatically brings with it the truth of ψ. For example, the conclusion (3) follows from premisses (1) and (2), because anyone who agrees with both premisses must also agree with the conclusion:

(1) $\forall x(F(x) \rightarrow G(x))$
(2) $F(a)$ (where a names an object in the domain)
(3) $G(a)$

To try it out, interpret F and G as above, and let a be the number 8; since the two premisses are true in this case (8 is divisible by 4), it follows that 8 is even. You can also let a be the number 9, but this time premiss (2) is false, so the argument does not come off the ground and the conclusion that 9 is even cannot be drawn. If you prefer an example without numbers, you can read $F(x)$ as 'x is a mammal' and $G(x)$ as 'x has lungs', for example, with $a = you$. If both premises are true—which you are perhaps in the best position to tell—then the conclusion, that you have lungs, must be true likewise. Logic is sometimes called *topic neutral*: Nothing hinges on how you read F, G, and a; if an argument is valid then its conclusion follows as long as the premises are true, regardless of what they are about. The 'follows from' relation is also known as *logical consequence*.

Various techniques have been invented for demonstrating that a conclusion *follows from* a given set of premises. One such technique rests on the use of combinations of simple *inference rules*, such as *modus ponens*:

Inference rule of modus ponens:

(1) $p \rightarrow q$
(2) p
(3) Therefore q

As before, the letters p and q are placeholders for arbitrary propositions. Modus ponens allows us, for example, to combine such premises as '23 is a prime number $\rightarrow$ 23 cannot be divided by 7' and '23 is a prime number', to yield the conclusion '23 cannot be divided by 7'. Likewise, when applied to the premises Short(150) $\rightarrow$ Short(151) and Short(150), modus ponens yields the conclusion that Short(151). Based on the definition of the conditional, it is easy to see that modus ponens is a *valid* rule, in the sense that it can never get us from premises that are true to a conclusion that is false. (It is a good exercise to try to invent other inference rules, and test their validity.)

To fully explain even the basics of predicate logic would require considerable space. Fortunately, the subject matter of this book does not require this. The version of the sorites paradox on which we focus relies on a long series of modus ponens applications, and little more than that. For our purposes, therefore, the summary given above suffices.

Vagueness can enter logic in different ways. One could, for example, model the behaviour of quantifiers such as 'almost all'. Perhaps

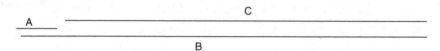

FIG. 11 Almost all As are Bs; almost all Bs are Cs; yet none of the As are Cs

surprisingly, this new quantifier behaves rather differently from its more uncompromising cousin 'all'. It is part and parcel of the meaning of 'all' (∀) that premisses 'all A are B' and 'all B are C' warrant the conclusion 'all A are C'. But if we replace 'all' by '*almost* all' (allowing, for example, at most 10 per cent of A as exceptions) then the conclusion no longer follows; in fact, it is possible[2] that *no A at all* are C (see Fig. 11).

Since the sorites paradox happens to hinge on predicates, the following sections will focus on deviations from classical logic where only the *predicates* in the formulas are vague. Yet this one difference will trigger others in its wake: some logics use truth values other than *true* and *false*, for instance; others redefine the conditional and the conjunction operator; some have ended up meddling with the law of the excluded middle. Many researchers, finally, have proposed deviations from the *logical consequence* relation of classical logic; modifications of this last type will not be discussed in detail, since they can be thought of as flowing from other, more fundamental deviations.

First deviation: Supervaluations and partial logic

How has classical logic[3] been modified to explain the sorites paradox? We start with an approach that is somewhat similar to the epistemicist's story. The epistemicist said: somewhere, at some unknown point, there is an exact threshold separating short from not short. Here we discuss an approach that views more or less *any* threshold as valid and asks whether there are things that must hold regardless of what threshold is chosen. The core logical concepts underlying this approach are those of a partial model[4] and of a supervaluation.

Supervaluations were first proposed by the philosopher Bas van Fraassen and applied to sorites by a number of authors, and most notably in a beautifully transparent research article by Kit Fine. Fine starts his exploration by focusing on the connection between such words as 'red' and 'pink'. Even if we do not know where the boundary between the two colours lies, Fine assumes that we do know that the two colours border on each other, so a blob of well-mixed paint cannot be both red and pink. Red and pink are related concepts. Given that they are vague at the same time, Fine says that there is a *penumbral connection* between them. To explain the concept of penumbral connections, he writes:

> They concern the common borderline cases of different predicates: if the blob is to be red it is not to be pink; if ceremonies are to be games, then so are rituals; if sociology is to be a science then so is psychology. Thus penumbral connection results in a web that stretches across the whole of language. The language itself must grow like a balloon, with the expansion of each part pulling the others parts into shape.

In order to do justice to penumbral connections between words, Fine adopts supervaluations. The basic idea of a supervaluation is simple: if you are uncertain about something, you may still be certain about anything that does not depend on the way in which your uncertainties are resolved. In other words, you call something true if and only if it is true under *all* the different ways in which your uncertainties might be resolved. A blob cannot be both red and pink because, regardless of how we define these colours, they do not overlap. More generally, 'a vague sentence is true if and only if it is true for all ways of making it completely precise'.

It is worth reflecting on this idea. Suppose you are asked how many people will come to your party, and most people have told you whether they will come. Your only source of uncertainty is that either John Doe comes to the party or Mary Doe does, but not both, since one of them will look after the little Does. In this situation, supervaluations say that

162

you do know *how many* people will come, because, regardless of how your uncertainty is resolved, the Doe family will delegate exactly one person to the party. Supervaluations do not allow you to resolve uncertainties in other ways. If you don't know the above-mentioned family arrangements of the Does, for instance, and all you know is that John's *probability* of coming is 50 per cent, and so is Mary's, then supervaluations do not allow you to estimate that one of the Does will be coming, and buy refreshments based on this estimate, since this is not certain. Supervaluations offer a cautious perspective on uncertainty, in other words. The idea of applying supervaluations to vague predicates is to regard the borderline cases of a vague predicate as a source of uncertainty, and to consider all the different ways in which these predicates can be made precise. So if a colour is somewhere in the border region between pink and red, then supervaluations allow you to say with certainty that the blob is either pink or red but not both.

How can this idea be grafted onto an otherwise classical logic? A natural picture arises if we assume an arbiter who told us that some particular people counted as short, and some others as not short, while possibly holding fire over others, because they are borderline cases. In logic, such an arbiter is called a *model*. If the model makes a decision about each conceivable person then it is called a *complete model*, otherwise it is a *partial model*. A little symbolism will come in handy. M will stand for a (partial or complete) model, and we shall write $M \models Short(a)$ to say that M has decided that a counts as short, while writing $M \models \neg Short(b)$ to say that M has decided b is not short. The notion of admissibility, discussed in the previous chapter, is now extended to all models, whether they are complete or partial:

Admissibility of a model M:
If $M \models Short(a)$ and $Height(b) < Height(a)$ then $M \models Short(b)$.
If $M \models \neg Short(a)$ and $Height(b) > Height(a)$ then $M \models \neg Short(b)$.

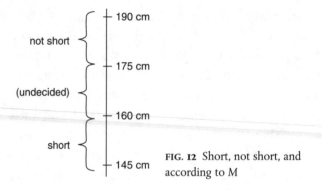

FIG. 12 Short, not short, and
according to M

Suppose the arbiter M decides that everyone below 160 cm is short, while everyone above 175 cm is not. If no other decisions have been taken then the following is a description of the arbiter's behaviour:

M ⊨ Short(a) if Height(a) < 160 cm, and
M ⊨ ¬Short(a) if Height(a) > 175 cm, and
neither M ⊨ Short(a) nor M ⊨ ¬Short(a) otherwise.

M's decisions regarding a set of people whose heights lie between 145 cm and 190 cm are pictured in Fig. 12.

The gap between 160 and 175 cm is called a *truth value gap*. M can be filled in many different ways, by adding decisions to the ones already made: the resulting, more opinionated models are known as extensions or *precisifications* of M. Of particular interest are precisifications that turn M into a complete model: these are supervaluations. Truth under supervaluations is called *supertruth*, and can be defined more precisely than before:

> *Supervaluational truth*: A statement p is supertrue with respect to a partial model M if and only if p is true in all precisifications of M that are complete. A statement p is superfalse with respect to a partial model M if and only if p is false in all precisifications of M that are complete.

Supertruth upholds the main logical laws. Consider, for example, 'John is either short or John is not short', an example of the law of excluded middle. Supervaluations treat this as just another penumbral connection. The sentence is supertrue because it is true regardless of where the threshold lies: depending on where this is, John may be short (making the sentence 'John is short' true) or not (making 'John is not short' true). In a similar way, one can easily show that contradictions (i.e. statements of the form p & $\neg p$) come out superfalse. This classical behaviour is one of the attractions of the supervaluational approach. Fine proves several other attractive properties for this approach. For example, once a sentence has received a truth value (e.g. it is supertrue given a certain incomplete model) then this truth value cannot be changed by precisifying the model.

How well do supervaluations perform in terms of explaining the sorites paradox? As before, we call the leftmost part of a conditional statement the antecedent, and the rightmost part the consequent. If all the listed conditionals were true then this would make it difficult to avoid the paradoxical conclusion that people taller than 175 cm are short. Assume, in the situation involving the partial model M sketched above, that a height difference of 1 cm is just below what is perceptible from a given distance. We abbreviate 'x is short' as Short(x):

Short(150) $\rightarrow$ Short(151)

...

Short(159) $\rightarrow$ Short(160)

...

Short(175) $\rightarrow$ Short (176)

Given the partial model M, many of these conditions are true. Consider the first one, for example: its antecedent and consequent are both true. In other words, someone who reasoned 'Short(150), so this conditional allows me to conclude that Short(151)' would not go wrong just yet. The conditional is therefore true. The same holds for the second conditional, and so on until one gets to

Short(159) $\rightarrow$ Short(160)

The antecedent of this conditional is true in M, because this is one of the decisions that this incomplete model does take. But the consequent can be either true or false: M does not say. Suppose someone reasoned 'Short (159), so this conditional allows me to conclude that Short(160)'. This argument would lead the reasoner from a true premiss, Short(159), to a conclusion that is neither true nor false, because of the incompleteness of M. (If M is completed in such a way as to make the consequent true then the conditional is true, but otherwise it is false.) For this reason, the conditional itself is neither (super)true nor (super)false in M. Viewed as a long conjunction of conditionals, the crucial premiss of sorites is (super) false. This is because, regardless of which complete precisification of M is chosen, one of the conditionals in the conjunction must be false. In ordinary language: wherever the threshold between short and not short lies, there the conditional will fail. This means that one of the premisses of the sorites argument is superfalse, and therefore the paradoxical argument does not come off the ground. A logical argument can force you to accept its conclusion only if it starts from true premisses.

The supervaluational account offers an elegant analysis of sorites based on the well-understood ideas of partial logic and supervaluations. Moreover, the account is sophisticated enough to acknowledge that vague concepts allow borderline cases, and it allows us to understand penumbral connections too. On the other hand, one could argue that the account does not do full justice to the vagueness of the predicates involved: the *one* crisp threshold of classical logic has been replaced by *two* crisp thresholds: one between short and the borderline area, and between the borderline area and not short.[5] Because these thresholds are crisp, they will sometimes separate objects which the perceiver is unable to tell apart. In other words, it would be hard to argue that Kamp's tolerance principle is satisfied. As a reminder:

> *Tolerance principle*: Suppose the objects *a* and *b* are observationally indistinguishable in the respects relevant to *P*; then either *a* and *b* both satisfy *P* or else neither of them does.

It seems then that supervaluations offer a good mechanism for reasoning about uncertainty, but a less good one for reasoning about observation predicates. Essentially, supervaluations flout the tolerance principle in the same way as epistemicism.

Second deviation: Context-aware reasoning

Context is vital for the interpretation of vague expressions. If I asked you, for example, whether a minute is a long time, you would probably not be able to give a confident answer until I informed you of the context of my question. If your heart fails to beat for a minute then that's a very long time; but a minute is hardly a long time in which to eat dinner, or to get your night's sleep. The expression 'a long time' is useful in a great variety of situations precisely because of this unbounded adaptability to the context in which the expression is used. The same is true for vague expressions more generally. To illustrate the point, let us look at what are some of the most widely applicable vague words: 'large' and 'small'.

Suppose 'large' had its meaning fixed once and for all, in terms of a fixed size. This would mean that the word could never meaningfully be applied to cows *and* to ants, but at most to one of the two species. For if the word was able to discriminate among cows (counting only some cows large), then it would fail to discriminate among ants, since all ants would be on the side of the smaller cows. Luckily, this is not how vague words work: a 'large' cow is large for a cow, and a 'large' ant is large for an ant. *Context-dependent* interpretation makes 'large' applicable to ants and cows, as well as to virtually everything else.

None of the responses to the sorites paradox discussed so far, however, has taken context into account. When we wrote 'Short(150)', for example, we did not say what other individuals we were comparing this individual with. The reason behind this omission was that, throughout the paradoxical argument, the context was seen as constant. We could, for example, agree to understand the supervaluational account of the paradox as if it was set in a classroom, with the pupils as the comparison class against which a person's tallness is judged. Nothing would change in the argument. So, although vague expressions are undeniably dependent on features of the context in which they are uttered, perhaps this is not relevant to the problem at hand.

Some responses to the paradox have, however, taken context as the linchpin of their solution, and after a slow start in the early 1980s, these responses are now enjoying considerable popularity.[6] The starting point for all these responses is to challenge the assumption that context must necessarily remain constant throughout the entire sorites argument, and to explore how context might change after each premiss, as a result of the fact that the premiss is uttered. The idea to let context change dynamically comes from years of linguistic research, in which many phenomena in natural language have yielded to such a view.

The archetypal example of an expression whose interpretation depends on context is personal pronouns such as 'he' and 'they', which can be interpreted only if you know the context in which they are uttered. The last word 'they' in the previous sentence, for example, refers to certain types of pronouns. If I decided to copy this word and paste it into a different part of the book, then it would instantly become dissociated from this interpretation and (if I'm lucky) take on a new one.

One might think that context-dependence is particular to personal pronouns, but nothing could be further from the truth. Longer expressions of the form 'the so-and-so' are equally dependent on context for their interpretation: consider an expression such as 'the queen': said in a film on Queen Victoria it refers to a historical figure, but said in a book about the present Queen Elizabeth, it refers to a living person. And even if we say something like '*everyone* knows so-and-so', we are unlikely to mean every living person: it is more likely that we refer to everyone in some situationally determined domain of individuals, such as an office or an aeroplane for example ('everyone knows that the plane is flying through severe weather'). All these phenomena, and many more, involve dependency on context. Detailed formal theories have been offered of many of them, and these have become so dominant that commentators have spoken of a 'dynamic turn' in linguistics.

One of the earliest and most influential representatives of the dynamic turn was Hans Kamp. After years of working in Richard Montague's tradition (see Chapter 6), his attention started to focus on the role of context in language interpretation. This shift in thinking would later lead to important new developments[7] in the study of language and meaning more generally. His ideas about vagueness and the sorites paradox, however, can be appreciated independently of these developments.

Incoherent contexts: Kamp's analysis

Kamp's account of the sorites paradox is a good illustration of the kind of flexibility shown by logicians in recent years. Far from seeing logic as an immutable inheritance, they view it as capital to be invested and diversified. Let's see how this works.

Kamp's starting point was the observation that we are *normally* willing to infer the shortness of an individual x from the shortness of a slightly shorter individual y, but not if it happens in a context where this leads to paradox; a context, in other words, where some object z which is indistinguishable from x has been called not short. In situations like this, x is caught in the middle, like a dog that doesn't know which of its owners to follow if they walk off in different directions. Such a situation is depicted in Fig. 13. The decisions embodied in the diagram can be seen as a precisification of the model M of Fig. 12.

Consider the paradox again, with all its premises of the form Short(x) → Short(x + 1). Kamp asserts that *each* of these premises is true taken separately but that, taken together, they are false. To explain how these two claims—which seem difficult to reconcile at first glance—fall out of his analysis, we need to talk about the notion of an incoherent context.

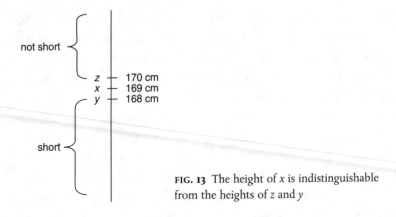

FIG. 13 The height of *x* is indistinguishable from the heights of *z* and *y*

And to do that, we need to uncover at least a glimpse of the formal system that Kamp used. I will simplify here and there, to allow its most important features to shine.

In the reality of actual discourse, contexts are complex situations, in which lots of different things are going on. But science almost invariably involves abstraction, so formal theories of context are typically much simpler. Kamp's theory is no exception. For him, a context is a bit like a (possible very small) partial model. More precisely, a context is a set of statements of the form Short(*x*). An example is the set $C_1 = \{$Short (150), ¬Short(151)$\}$, which represents the situation in which you have decided to call the individual whose height is 150 cm short, and the individual whose height is 151 cm not short. To make his logic work, Kamp needs a new concept, for which he uses the word *coherent*. Whether C_1 is coherent depends on one's powers of perception. If a difference of 1 cm is not enough to be distinguishable, then the above-mentioned context C_1 is incoherent. More generally, Kamp defines coherence as follows:

> A context is incoherent if and only if it contains statements Short(*x*) and ¬Short(*y*), where *x* and *y* are indistinguishable in terms of height.

170

This is like the tolerance principle but with a contextual twist. Turning to the paradox, Kamp first focuses on each of the above-mentioned premisses asking, for each of them, whether it is true in the *relevant* context, which consists of all the decisions taken so far. Suppose these decisions are that 150 cm is short and 175 cm is not. Our context, in other words, is the collection {Short(150), ¬Short(175)}. What, under these circumstances, are we to make of a conditional such as

Short(159) → Short(160)?

To answer this question, we have to know how Kamp defines the meaning of conditional propositions. Here Kamp deviates from standard practice, taking his lead from sentences such as 'If John comes home early then he is in a good mood,' containing the anaphoric word 'he'. We can interpret this word only because we have read about John earlier on in the sentence. Based on examples like this, linguists know that the consequent of a conditional statement should be interpreted in the light of the antecedent. Kamp applies this idea to sorites sentences. Let's see how.

Suppose we are asked to judge the truth of the sentence 'Short(159) → Short(160)' in some prior context C. Whether this statement is true must have something to do with the truth of its two parts, 'Short(159)' and 'Short(160)', but how exactly? Kamp's proposal is that the truth value of the statement 'Short(160)' is judged not in C, but in the context that results when the statement 'Short(159)' has been added to C. In our case, this means that we are judging the truth of 'Short(160)' in the context {Short(150), Short(159), ¬Short(175)}.

But what does the context {Short(150), Short(159), ¬Short(175)} make of 'Short(160)', given that this statement itself is not a member of the context? In Kamp's system, the fact that 159 has been judged to be short means that this context has, as it were, *implicitly* decided that 160 is short as well, because the two are so similar. Kamp accomplishes this by stipulating that

> A statement of the form Short(*a*) is true in a context *C* if *C* contains a statement of the form Short(*b*), where *b* is either larger than *a* or indistinguishable from *a*.

The effect is that Short(160) must be true in a context that contains Short (159). This does not mean, however, that any paradoxical conclusions follow, for Kamp has another trick up his sleeve: although each of these conditions is true individually, they are collectively false. How is this done?

When we are talking about the conditionals 'collectively', we mean their logical conjunction, that is, the statement saying that they are all true. Abbreviating the first of the statements as p_{150} and the last one as p_{175}, we are talking about the statement

$$p_{150} \ \& \ p_{151} \ \& \ \ldots \ \& \ p_{175} \ \& \ p_{176}$$

Kamp's reason for calling this conjunction false rests on the notion of an *incoherent* context that was defined above. He defines a conjunction to be true only if it does not make a context incoherent:

> *Truth of conjunctions*: A statement of the form A & B is true in a coherent context C if A is true in C and B is true in C and the context that results from adding A and B to C is also coherent.

In the cases of interest to us, A and B are sorites conditionals. The idea here is that, when you add a conditional to a context, you also add any facts that follow from it. For example, when adding the conditional Short(159) → Short(160) to a context that already contains Short(159), you have to add Short(160) as well. This means that, when *all* the sorites conditionals are added to the original context {Short(150), ¬Short(175)}, the result is incoherent, for example because it contains Short(174) as well as ¬Short(175). Given the proximity of Mr 174 and Mr 175, this represents the kind of clash that is, for Kamp, the hallmark of incoherence. Given the clause for the truth of conjunctions, this makes the conjunction of conditionals false. In other words: looking at the many conditionals that make up the crucial premiss, each of them is true in the original context, but they are false collectively, in this particular context, because they make it incoherent.

Leaving further technical details aside, the structure of Kamp's solution is undeniably attractive. After all, most of us would be willing to accept the truth of a statement such as Short(159) → Short(160), as long as many similar statements *together* didn't have such awkward consequences. It is this fundamental intuition that Kamp's account captures. I believe that this is a genuine insight into the way people use vague expressions and it should therefore, ideally, be reflected in a theory of vagueness. Yet, Kamp's approach is not without its drawbacks. For a start, it might be argued that Kamp's mechanism removes him uncomfortably far from classical logic, given that so much of the mechanism deviates from it. Moreover, Kamp's theory shares an important problem with the supervaluational account: it predicts that, given the right setting, *arbitrarily similar* objects can be told apart. Kamp's approach does not obey the tolerance principle, in other words. For reasons of this kind, researchers have tried to take these ideas in new directions.

Context and indistinguishability

The philosophers Nelson Goodman and Michael Dummett had some interesting ideas, with roots in the works of Bertrand Russell, concerning the way in which indirect evidence can sometimes be used to tease two objects apart. Let us see how these ideas shed light on the sorites paradox.

Imagine an experiment in which you are exposed to the poetically named Mr 150, and asked whether this person is short, to which you say 'yes'. Next, you are shown Mr 151, who is just one centimetre taller, and asked whether you can distinguish this person's height from that of the previous one. Let us suppose that you cannot. Then you are asked whether Mr 151 is short, and presumably you will say that he is; after all, you have just admitted that Mr 151 is indistinguishable from Mr 150! The procedure goes on and on: you are always asked to compare pairs of people: Mr 151 and Mr 152, then Mr 152 and Mr 153, and so on until the people to be judged are proper giants.

So far so good—or rather, so bad, because a paradox is derived in the usual way. The philosopher-logicians Frank Veltman and Reinhard Muskens appear to have been the first to see the capacity of Goodman and Dummet's ideas to shed light on the sorites paradox. Their take on the situation is to assume that all the people whose height you have judged in the experiment will linger: no one walks away from the scene. Now the idea is that the context of the indistinguishability judgement at the heart of this experiment ('Do you see a difference between these people's heights?') is subtly different each time: the first time, all you have to work with is the two people you are comparing. The second time, you are looking at three people: the two you are comparing (Mr 151 and Mr 152), plus the very first person (Mr 150). You have more and more information at your disposal each time. This may very well matter to your distinguishability judgement. Let us look further.

When two people, x and y, are indistinguishable in height, we write $x \sim y$. Goodman and Dummett argued that x and y might *become* distinguishable by virtue of a suitably chosen third person z. For example, if x is slightly taller than y, then a third object z might be found such that z is visibly taller than y while not being visibly taller than x. If this happens, z can help the observer to realise that x is taller than y. In other words, it depends on the presence of a *help element* such as z whether x and y can be distinguished from each other. In a context where a help element is present, x and y are *no longer* indistinguishable (see Fig. 14).

Veltman and Muskens's account is couched in terms of symbolic logic, but its essence can be conveyed without much formalism. The main idea is to agree that the sentences that make up the paradox should be judged in the context *in which they appear*. This is done in such a way that the elements of the first pair (Mr 150, Mr 151) are compared in a context where these objects alone are present. As we have seen, this makes the two indistinguishable. But now consider the second pair (Mr 151, Mr 152). The context in which they are compared contains not only 151 and 152

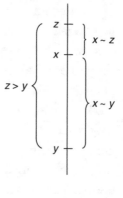

FIG. 14 The height of z helps the observer to
distinguish between the heights of x and y;
similarly, y helps to distinguish
between z and x

themselves, but the old 150 as well, which is stored in memory so to
speak. If, as we have assumed, 150 and 152 *are* distinguishable from each
other, then this difference between 150 (which is distinguishable from 152)
and 151 (which is not) allows the observer to infer that 151 and 152
must have different heights even though he cannot directly see it
(see Fig. 14, with y in the role of 150).

The upshot is that, once the sorites conditionals are understood
properly, they do not take you beyond a certain point in the sorites
sequence, since at some point you won't be able to find objects that
are indistinguishable from the previous ones. Although the Goodman
and Dummett construction helps to give substance to it, the argument
does not stand or fall with this construction. It seems plausible that
objects are easier to distinguish because the observer has seen more
other objects, and any account of this phenomenon could potentially
be used to explain the sorites paradox along the lines sketched above.
The crucial idea, one might say, is to give distinguishability a *memory*.

This type of account can be developed in a number of ways. For
example, as I have shown elsewhere, it can support the view that sorites
hinges on a subtle equivocation in the words that make up the argument.[8] This idea will be illustrated later in this chapter, in connection
with a new proposal.

175

Veltman and Muskens's approach can be modified in various other ways. One can, for example, allow it to take memory limitations into account: sometimes a help element such as Mr 150 may be too far back in the past to be taken into account when we try to see a difference between Mr 151 and Mr 152. By building in a proper notion of forgetting into the model it may become possible to explain, for example, why gradual changes tend to go unnoticed when they occur very slowly: if this happens, the memories that could have told us that something is changing may simply have been forgotten. An approach of this kind could also explain why alertness matters: someone who pays close attention and commits facts to memory will stand a better chance of noticing the change.

Although these analyses have considerable strength, I now believe that approaches of this kind leave certain important things to be desired. For one thing, they fail to explain what happens in the original version of the experiment, where help elements are carefully excised from the picture, because people were judged in pairs. Furthermore, these approaches treat indistinguishability as if this notion itself was crisp. But crisp models of indistinguishability are ultimately untenable, as we shall see, and this will force us to consider new ways of thinking about the paradox.

The trouble with variety

The tolerance principle stated that when two stimuli are sufficiently similar, they should be judged in exactly the same way. On reflection, this is unrealistic. As any observer of people knows, slightly erratic behaviour is the norm. Psychophysicists—who build mathematical models of human perception based on controlled experiments—have known this for at least a century, but semanticists have only recently come to terms with these ideas, thanks to the work of Delia Graff Fara and others. The point is worth emphasizing by means of a thought experiment, even at the risk of belabouring the point.

Suppose you let a human subject estimate the heights of a set of people. Suppose, furthermore, that 100 people of exactly the same size were presented to the subject at irregular intervals. (To break the monotony, you should also include people of different heights, but the subject's responses to these will not matter.) All this should be done in such a way—perhaps by showing people's silhouettes—that the subject had no other way of identifying the people they are judging. To make direct comparisons impossible, people should be judged one by one, and the subject should be asked to express her estimates as a whole number of centimetres. It is then extremely unlikely that the same estimated height will result all 100 times. A much more typical outcome would show a probability distribution in which some values appear more often than others, followed by other values that occur slightly less often, which in turn are followed by yet others that occur even less often, and so on. If you are lucky, the *correct* height is mentioned most often, but there is no guarantee that this will happen.

Why this lack of uniformity? We are creatures of flesh and blood, whose brains are not always in the same state. Both men and women are subject to various cycles, for example involving day and night. We can be tense or relaxed, tired or rested, happy or unhappy, in constantly varying combinations and degrees. Clearly then, the tolerance principle is on shaky ground. For if people of the same height are judged differently from time to time, then the same must surely be true of people whose heights are just *similar*. The tolerance principle forbids us to distinguish between stimuli that resemble each other closely, yet we routinely distinguish even between stimuli that are *identical*.

Another important concept loses its innocence as well when variety is taken into account: the notion of a *just noticeable difference* (JND). A JND is the smallest difference that someone can notice. Consider the following argument which we might call the paradox of perfect perception. Suppose we are testing a subject's sensitivity to the loudness (i.e. amplitude) of sounds. The subject is not deaf, and his JND is known. Then the following paradox can be construed:[9]

The paradox of perfect perception (PPP):

1. Some sounds are so similar that the difference in amplitude between them cannot be perceived by any human ear.
2. Given *any* two sounds A and B, where B is louder than A, there exists a third sound C, louder than A and B, such that the subject can hear that C is louder than A, but not that C is louder than B.
3. Therefore, when confronted with A, B, and C, the subject can infer that B is louder than A.
4. Since the difference in loudness between A and B can be arbitrarily small, it follows that the subject can perceive arbitrarily small differences in amplitude, contradicting (1).

An analogous argument involving people's heights is easy to construct. Embarrassingly, none of the theories discussed so far has anything to say about the PPP and, to the extent that they buy into a crisp notion of distinguishability, they are in a poor position to find a solution to it.

Variability is not a feature of human perception alone, since each measurement tool is subject to a certain amount of variation. We have seen this in the previous chapter, when discussing the margin of error associated with a weighing balance. One might also think of an hourglass: the grains of sand are not always positioned in the same way, causing tiny variations in the time they need to make their way downwards through the neck of the instrument; air humidity and temperature have a role to play as well, and because individual grains of sand can stick to the glass, there are slight uncertainties in determining when the upper chamber is empty. As measurement instruments become more sophisticated, the variations involved in measurement are diminished further and further, but they cannot be eliminated.

Third deviation: Introspective agents

To do justice to individual variation, let us use a robot as a caricature of a person. This will allow us to explore different assumptions about the way our senses work. The caricature is inspired by a formal model proposed by the theoretical computer scientist Joseph Halpern, although my use of it for explaining sorites will be quite different from his. The protagonist of the story is a robot that is able to measure the height of a person, but only with some limited, precisely stable accuracy.

Two robots may differ in many ways; they may have different sensors, for example. Particularly relevant for us are the perceived height of a person, and the *margin of error* associated with the robot. Robots may also differ in terms of the thresholds they use when making decisions. One robot might call anyone short whose height does not reach 160 cm, while another might set this standard at 155. Crucially, a robot does not *know* the person's height, except indirectly and fallibly, through its own observations. Moreover, a robot's observation may produce different readings at different times. Strictly speaking, this dependence on time (or on the state in which a robot finds itself, in Halpern's proposal) should be reflected in our formalism. We shall, however, simplify by suppressing these subtleties. Suppose a robot measures the height of *p* at a particular moment in time, and the result of this measurement is reflected in the value of 'perceived height of *p*'. Suppose Robo and Boro can be characterized as follows:

Robo: perceived height of *p*   = 154 cm
      margin of error   = +/− 5 cm
      lower threshold for being short   = 160 cm
Boro: perceived height of *p*   = 153 cm
      margin of error   = +/− 3 cm
      lower threshold for being short   = 155 cm

We do not care where these values come from. For all we know, a robot's threshold might derive from a comparison with its own height, in which case Robo must be taller than Boro, or from the height of a shelf on which a lethal weapon is stored: all we need to know is the robot's threshold.

How will each robot assess p's height? We distinguish between two kinds of assessment: an immediate one, and one involving reflection. For the former, we shall use the word 'appear' (as in 'this person *appears* to me to be short'), for the latter the word 'certain(ly)' (as in 'according to my standards, this person is *certainly* short'). The choice of words is immaterial, since they have no meaning other than the defined ones, which are related to the specification of a robot, as we will explain.

Let us assume that each robot is aware of its own limitations: they are introspective, as we say. Now Robo might reason as follows: p *appears* short, since his perceived height falls short of the threshold below which Robo regards someone as short. In fact, Robo can be *certain* that p is short, since Robo's margin of error does not exceed the difference between p's perceived height and the threshold employed by Robo. More generally, p will appear short to a robot if, in the robot's specification, p's perceived height falls below the threshold value; the robot can decide that p is *certainly* short if and only if p's perceived height plus the margin of error falls below the threshold. Using Height(x) to denote the perceived height of x, and MoE to denote the Margin of Error (Fig. 15):

x *appears short* if and only if Height(x) < threshold.

x is *certainly short* if and only if Height(x) + MoE < threshold.

Even though p appears shorter to Boro than to Robo, and Boro is better at estimating heights, Boro cannot be certain that p is short, because it sets stricter standards for shortness than Robo. Our robotic caricature can be refined in various ways. But even our simple model allows us to make some interesting observations. Suppose, for example, we define indistinguishability (represented by the symbol $\sim$ as before) as follows,

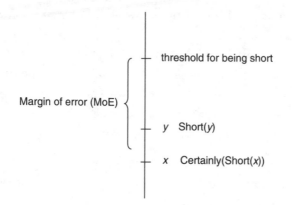

FIG. 15 *x* is certainly short, but *y* is just short

using Height to abbreviate the height of a person as measured by one particular robot (using double bars to denote absolute value):

$a \sim b$ if and only if $\|\text{Height}(a) - \text{Height}(b)\| < \text{MoE}$

In other words, *a* and *b* count as indistinguishable if their height difference, as measured by that robot, is smaller than the margin of error (MoE). The real difference between them could be anything up to twice the MoE, because each of the two can vary both upwards and downwards. Similarly, if a robot perceives *p* to be taller than *q* (i.e. Height(*p*) > Height(*q*)) then *q* may actually be taller than *p*, even if the perceived difference between *p* and *q* exceeds the MoE, because both measurements may be subject to error.

What would be a good way to render the tolerance principle? Several alternatives come to mind, all of which involve two objects that stand in the just-defined relation $\sim$. Suppose *x* is perceived on one occasion, and *y* on another, and the difference between these two readings amounts to something less than the MoE. Suppose *x* appears to be short. Now it does *not* follow that *y* appears short as well, since the MoE may easily take *y* above the threshold *m*. The most obvious version of tolerance does not hold, therefore. Yet, a version of tolerance can be salvaged if we assume that *x*'s height is perceived to be so low that x's real height must be below m even when the MoE is taken into account.

If Certainly(Short(x)) holds while $x \sim y$ is also true, then certain inferences can be made. Admittedly we cannot conclude that Certainly (Short(y))—as is easy to verify—but we can conclude that Short(y), because y must lie below the threshold m. Summing up, (a) and (b) are invalid versions of tolerance, but (c) is valid:

(a) If Short(x) and $x \sim y$ then Short(y).
(b) If Certainly(Short(x)) and $x \sim y$ then Certainly(Short(y)).
(c) If Certainly(Short(x)) and $x \sim y$ then Short(y).   (*Valid*)

These observations can be applied to sorites in a straightforward way. Suppose our inanimate friend Robo observes that a person x, whose height it estimates to be 154 cm, is certainly short, because this estimate lies more than one MoE below his threshold. Now Robo is confronted with a person y, whose height it estimates to be 157 cm. Using (c), Robo can infer that y must *appear* short to it *on this occasion*. The stronger conclusion, that Certainly(Short(y)), does not follow. Had Certainly (Short(y)) followed, a lethal sorites chain would have been set in motion, but the valid version of tolerance does not allow Robo to make any inferences that take it further than its margin of error allows.[10]

The model based on introspective agents, which is presented here only in outline, shows how non-trivial inferences can be made from one observation to the next: if we observe someone to be well below the threshold for shortness, for example, then individuals of similar height must at least be categorizable as short. By qualifying perception judgements in this way, it becomes possible to do justice to *degrees* in a logic that is close to classical logic in many ways. Additionally, the model represents a first attempt to take the *variability* of perception judgements into account.

Something else is worth remarking about this model, namely the role that *ambiguity* plays in it. If we examine its logic in more detail then a pattern emerges to which we have alluded earlier in this chapter: when the variability of perception is taken into account, tolerance can be understood in different ways: it can be understood in a perfectly sensible way (for example using (c) above), but then it is too harmless to derive a sorites argument; it can also be understood in other ways (such as (a) and (b)), which are powerful enough to build a sorites argument, but which are invalid. What this analysis suggests is that *ambiguity* is the culprit behind the paradox. We are taken in by the sorites paradox because its language confuses us: its crucial premisses cannot be denied, because they have an interpretation according to which they are true. Conversely, we cannot deny that the premisses are paradoxical—because they have an interpretation according to which they lead to paradox. We are safe from reaching paradoxical conclusions because there is no interpretation that is true *and* leads to paradox.

To see how ambiguity can lead us astray, here is a comparison that involves a very different type of domain, and different words. Suppose a politician, in a democratic country, argues in favour of referenda. Having exhausted all other arguments, she decides to appeal to principle, arguing as follows: 'Everyone is in favour of democracy; democracy requires that every citizen can vote; therefore referenda must be held.'

The cleverness of this argument lies in the ambiguity of the word 'democracy'. If the word is understood very generally, as denoting a *fair* political system, which treats every citizen justly, then the politician is right to assert that her audience believes in democracy (or so we assume). Let us call this uncontroversial idea democracy$_1$. But it is not obvious why democracy$_1$ (i.e. democracy in the sense of a fair system) requires referenda. Our politician has exploited the fact that the word 'democracy' has a more specific sense, denoting a political system entirely based on *majority rule*. Calling this highly specific concept democracy$_2$, it is plausible that belief in democracy$_2$ would force one to believe in referenda, but belief in democracy$_2$ is far from universal. To sum up: everyone believes in democracy$_1$, but this belief does not force us to believe in referenda; democracy$_2$, would force us to believe in referenda … but not everyone believes in it. The model based on introspective agents contends that the cogency of the sorites paradox rests on the same mechanism.

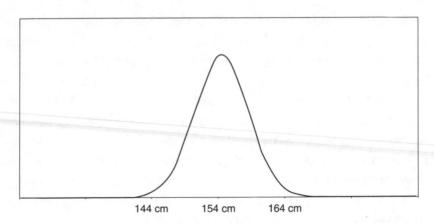

144 cm   154 cm   164 cm

FIG. 16 A distribution of height estimates for a person whose height is 154 cm

Doubts about errors

But nagging questions remain. To start with, it may be unrealistic to assume that perceivers are aware of their own margin of error. It is even more unrealistic to assume that the MoE is crisply defined. If our MoE is exactly 5 cm then an error of 5 cm is made out to be possible, whereas an ever so slightly larger error is ruled out. This is plain madness: it is surely impossible to ascertain that 5 cm, as opposed to 4.75 cm or 5.25 cm for example, is the right figure. If a large-scale experiment was conducted, and the frequency of errors of all different sizes was plotted we would find a perhaps broadly symmetric, roughly bell-shaped curve. The curve would show smaller errors to be more likely than larger ones (see Fig. 16).

What these considerations show is that once the variability of human perception is taken into account, such concepts as just noticeable difference, but also margin of error, should arguably become *probabilistic*. It would be technically possible to inject probabilities into the framework outlined in this section, but this would be rather artificial. If probabilities are to play a role, they deserve better than to be added as an afterthought.

There is another reason for changing course. We have devoted a lot of space to solutions to sorites that hinge on indistinguishability, but solutions of this kind have little to say about other versions of the

184

paradox. The paradox of the strict finitist (Chapter 5), which revolved around the concept of an acceptably small number, is a case in point: presumably we know the difference between any two natural numbers, so indistinguishability never holds. Or consider the concept of baldness: even if we had a simple method of counting the hairs on someone's head, 'bald' would remain a vague concept, perhaps because it is, in Fara Graff's apt turn of phrase, interest-relative: the more hairs, the more protection from cold and sunburn; and different people might be interested in protection to different degrees. In cases such as this, indistinguishability does not seem to be at the heart of the problem. What we do find is *degrees*, wherever we look. It is time we took them seriously.

Things to remember

◆ *Classical logic* is a sophisticated and well-understood mathematical theory that has been used to analyse many issues connected to the notion of a valid argument. Among other things, it is characterized by the fact that it regards each statement as either true or false, with nothing in between.

◆ The responses to the sorites paradox discussed in this chapter attempt to give vague expressions a place in a logical framework. They attempt to do this in such a way that the overall framework stays as close to classical logic as possible. Additionally, they try to do justice to the tolerance principle, which states that words such as 'tall' and 'short' cannot distinguish between objects that our senses cannot tell apart.

◆ Context-based approaches to sorites show us how vague state-ments are assessed in the light of earlier assessments. One such situation arises if a new assessment threatens to cause paradox because of earlier assessments. For example, Mr 172 may be about to be judged to be short, whereas an earlier assessment says that the indistinguishable Mr 173 is tall. Another such situation arises if an assessment involves an object that is distinguishable from previous objects only by virtue of help elements in the style of Goodman and Dummett.

◆ By adopting a model of *forgetting*, context-based approaches can help us understand why slow changes often go unnoticed.

◆ Explanations of the sorites paradox cannot afford to overlook the fact that people's understanding and use of vague expressions is variable. One important consequence of this fact is that distinguishability is not a crisp concept. An analysis of the *paradox of perfect perception* suggests the same conclusion. This conclusion diminishes the appeal of most context-based approaches, because they tend to rely on a crisp notion of distinguishability. Moreover, it forces one to consider weakened versions of the tolerance principle, as is done by the approach based on intro-spective agents and by the approaches in the following chapter.

◆ The idea that the crucial premiss of the sorites argument is *ambiguous* offers an attractive pattern for explaining why the argument is plausible and fallacious at the same time. This pattern underlies some context-based approaches and the agent-based approach as well.

9

Degrees of Truth

All traditional logic habitually assumes that precise symbols are being employed. It is therefore not applicable to terrestrial life but only to an imagined celestial existence.

RUSSELL (1923)

We have seen quite a few responses to the sorites paradox. Yet in one important respect, all these approaches are similar: when confronted with a claim of the form 'so-and-so is short', they can only assess the claim as true or as false unless, in some cases, they may be able to reserve judgement. Although these approaches afford considerable insight, they leave much to be explained. For one thing, we have seen that they represent key concepts (just noticeable difference, margin of error) as if they were crisp, thereby begging important questions. But there is more. If John is taller than Bill, then 'tall' applies *more* to John than to Bill, and we are more likely to call John tall. The approaches discussed so far have difficulty coming to terms with this intuition.

To explain the issue, here is a story: a tale of a stolen diamond. It is set in Beijing's Forbidden City, long ago. A diamond has been stolen from the Chinese Emperor and, security being tight in the palace, the thief must have been one of the Emperor's 1,000 eunuchs. A witness has seen a suspicious character sneaking away. He tried to catch him but failed, getting fatally injured in the process. The scoundrel escapes. With his last breath, the witness reports 'The thief is tall!', then gives up the ghost. How can the Emperor capitalize on these momentous last words?

Suppose the palace harbours a *classical logician*. After some deep thought, this logician decides that 'tall' needs to be interpreted as 'taller than average'. If the Emperor listens to him, his men will gather all those eunuchs who are taller than average, perhaps about 500 of them. Each of them has to be searched. This is a lot of work, and the procedure may alienate the eunuchs: never a good idea, if history is a guide. And matters could be worse. It is conceivable, after all, that the witness used a more relaxed notion of 'tall', and the thief was actually just *below* average size! In this case, the perpetrator will be overlooked when the tall eunuchs are gathered. Since classical logic does not make any height distinctions between non-tall eunuchs, all of them are essentially alike, so the Emperor's men turn to the eunuchs they consider not to be tall, searching them in arbitrary order. In the worst case all 1,000 eunuchs need to be searched. No doubt, the fate of our classical logician will be terrible.

But perhaps the palace logican is an adherent of *partial logic*. In this case he can do better, because of his ability to separate the eunuchs into three rather than two groups. Assume 100 eunuchs are definitely tall, 500 are definitely not tall, and 400 are doubtful. The eunuchs in the 'definitely tall' category are more likely to be called tall than the ones in the 'doubtful' category, while no one in the 'definitely not tall' category would ever be called tall. To put some arbitrary figures to it, let the chance of finding the thief in the group of 100 be 50 per cent (i.e. the probability is 0.5) and the chance of finding him in the group of 400 likewise. Under this scenario, it pays to search the 'definitely tall' eunuchs first, as one may easily verify.

But the Emperor never went to university. Being self-taught, he is likely to do much better than the classical and the partial logician. He will ask his men to order the eunuchs according to their heights. First the tallest eunuch will be searched, then the tallest but one, and so on, until the diamond is found. Under reasonable assumptions, this strategy is faster than each of the other ones, because the taller a person is, the more likely the witness is to have described him as tall. (The same advantage obtains as in the case of partial logic, but on a larger scale.) None of the theories that we have seen so far can explain why the Emperor's strategy makes sense, given that none of them models the *degree* to which a person is tall, or their rank order of heights, for that matter. To classical logic, for instance, all the people in the 'tall' category are essentially alike: the height differences between them are ignored.

But a model that can do justice to the Emperor's strategy is, in fact, available. The starting point is the hypothesis that *degrees* are at the core of vague concepts, and that degrees of *truth* need to be built into our logical apparatus. Not just two or three, but as many as are necessary for making all the necessary distinctions. We shall call the resulting theories *degree theories*. Degree theories have never been popular among theoreticians. I will argue that this is partly an accident of history. The accident I am referring to is the fact that the best-known degree theory is least well placed to shed light on logical arguments such as the sorites paradox. This theory is known as fuzzy logic. It's a bit as if the first vegetables you had ever tasted were Brussels sprouts: you didn't like them and decided not to eat all vegetables for the rest of your life. To stick with the metaphor, this chapter has two lessons: Brussels sprouts are rather good for you, and some other vegetables are even better.

Degrees of truth are the topic of this chapter. After a description of fuzzy logic and some of its applications, its shortcomings will be discussed. This discussion will lead us to a less well-known degree theory, which is based on the *probability* of a proposition being regarded as true. I will argue that these probabilistic theories have the potential to shed important light on vagueness. There will come a cost, however: our expulsion from Boole's two-valued paradise.

Fuzzy logic

Early in the previous chapter, a logical system called partial logic was outlined. Partial logic added one truth value to the two familiar truth values of classical logic, and this enabled it to model vagueness more subtly, though some problems remained. But if a division into three zones is too coarse-grained, then why not go further, by distinguishing an infinity of truth values? A proposal along these lines was made in the 1960s and 1970s by Lotfi Zadeh and others, based on work by Max Black (who, like Zadeh long after him, hailed from Baku in the then Soviet Republic of Azerbaijan) and the Polish logician Jan Łukasiewicz in the 1920s and 1930s.[1] They proposed a large range of truth values between 0 and 1, where 0 means 'completely false', 1 'completely true', and all other values are intermediate degrees of truth. Fuzzy logic has considerable potential for specifying the behaviour of computer programs, as we shall soon see, because it allows programmers to think about complex problems in common-sense terms. Fuzzy logic also appealed to a band of esoteric thinkers who saw in it the demise of Western science as we know it: too long had science simplified a complex reality by thinking about it in terms of truth and falsehood; it was time that we all embraced the ancient Oriental wisdom of fuzzy thinking. Let us see what fuzzy logic can do for us. We present the logic informally, referring the reader to the literature for more mathematical accounts.

The first step in the construction of fuzzy logic is to set up a continuum of truth values, ranging from 0 (entirely false) all the way to 1 (entirely true) with all the real numbers greater than 0 and smaller than 1 in between. A larger number means that something is true to a higher degree. 'Denmark is a large country', for example, would receive a truth value lower than that of 'Spain is a large country' which, in turn, would receive a lower value than 'China is a large country'. This, of course, is not saying very much yet. Suppose the second-largest country is half the size of the largest one, does this mean that one truth value is twice as large as the other? We shall later return to questions of this kind. Fuzzy logicians have typically been relaxed about these matters, emphasizing the flexibility of their framework.

Suppose you need a function that says, for each height that a person x may have, what the truth value of the statement 'x is tall' is. If you want to represent the intuition that all heights below a certain interval should count as not tall *at all*, while all those above the interval count as tall without qualification, then a function such as the following might be suitable (where v stands for degree of truth):

$v(\text{Tall}(x)) =$
0 if $x < 150$
1 if $x > 190$
$(x - 150) / (190 - 150)$ otherwise

This function consists of three intervals: two flat ones below 150 cm and above 190 cm, and a linear interval in between. Information of this kind is called a membership function, which is conveniently displayed in a membership graph (Fig. 17). Sometimes fuzzy logicians use smoother functions, without the sudden change between these three intervals. Although this is probably preferable in theory, diagrams with just three straight-line segments are often used, because they make it easier to perform calculations.

Fuzzy membership functions allow us to improve upon the crude dichotomy between crisp and vague concepts that is often used in order to simplify matters. Consider a concept such as 'being taller than 160 cm'. We have seen in Chapter 3 that this concept is not *entirely* crisp, because there can be individuals whose height is so close to 160 cm that even perfectly conducted measurements might differ on whether they are taller than 160 cm or not. Fuzzy logic can represent this situation very naturally by using a membership function that consists *almost* entirely of two flat parts (for 1 and 0) plus a very small area of intermediate values. In this way, fuzzy logic allows us to say that vagueness itself is a matter of degree: as a first approximation, for example, one might measure the *degree of vagueness* of a concept as the length of its intermediate segment (whose values are greater than 0 but smaller than 1) as a fraction of the width of the graph as a whole. Subtler versions of this idea are easily constructed.

The membership functions of different expressions are related to each other in interesting ways. It seems reasonable, for instance, to define

191

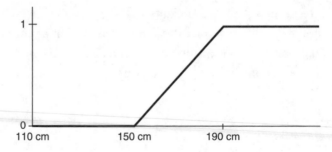

FIG. 17 A fuzzy membership function for the concept 'tall'

short as the converse of tall. One formula that has been used here defines the truth value of Short(x) as $1 - v(Tall(x))$: the taller you are, the less short. A substantial literature exists on how functions may be constructed for all kinds of different related expressions. A recent textbook on artificial intelligence offers these tentative definitions:[2]

$v(Very\ tall(x)) = (v(Tall(x)))^2$
$v(Extremely\ tall(x)) = (v(Tall(x)))^3$
$v(A\ little\ tall(x)) = (v(Tall(x)))^{1.3}$
$v(Slightly\ tall(x)) = (v(Tall(x)))^{1.7}$

It is worth playing with these definitions, to see how they pan out. Suppose, for example, that Billy is tall to a degree of 0.5; this would make him 'very tall' to degree 0.25 and 'extremely tall' to degree 0.125. The point of these definitions is not that they are exactly right, but that they are as good an approximation as money can buy. The same is true for the assignment of truth values to simple statements. Sometimes empirical work is invoked to defend the flat-line segments in the membership graph above, for example, while on other occasions the systematic connection between 'tall' and 'taller', for example, is emphasized:

x is taller than y if and only if $v(Tall(x)) > v(Tall(y))$.

(In words: being taller means being tall to a higher degree.) It is easy to see that this stipulation is not compatible with the flat-line segments in the membership function that was used above, because the stipulation

192

requires a membership function that assigns different membership values *for each height*. No one seems to care much, and the prevailing attitude is that anyone can define the membership function that she requires. Before discussing these issues further, let us turn to fuzzy logic's treatment of compound expressions involving words such as 'and', 'not', and the like, whose meaning is crucial when vague expressions enter logical argumentation. The name of the game is to redesign classical logic, as it were, this time making optimal use of the entire spectrum of values between 0 and 1. In doing so, the extremes of 0 and 1 need to be respected: on these two values, fuzzy logic should always produce the same outcomes as classical logic.

We start with negation, which is perhaps the least problematic of the lot. Statements involving 'not' ($\neg$) are usually given the *complement* of the value of whatever 'not' applies to: the truth value of a statement of the form $\neg p$ equals $1 - p$. Thus, if p has the value 0.6, then $\neg p$ has the value 0.4. If p has the value 1 then $\neg p$ has the value 0, just as usual. Note that this arrangement makes the negation of a statement different from the *opposite* of the statement: the negation of 'being extremely short' is not 'being extremely tall', but '*not* being extremely short'.

How about compound statements involving 'and' (&)? Fuzzy logic normally defines the truth value of a statement of the form $p \& q$ to be equal to the *minimum* of the value assigned to p and the value assigned to q: a chain of statements is only as strong as its weakest link. It is not obvious, however, that this is always reasonable. Combining these two definitions, for example, it follows that $Short(x) \& \neg Short(x)$ will often have a remarkably high truth value: if the value $Short(x)$ is 0.5, the value of $\neg Short(x)$ is $1 - 0.5$, which is also 0.5, hence the value of $Short(x) \& \neg Short(x)$ is 0.5 as well. What we are seeing here is that the law of non-contradiction no longer applies: contradictions cannot be relied on to have a value of 0 which is unusual, and probably undesirable.

But aren't we being unfair? In Chapter 7 we agreed to accept non-standard logics as a fact of life. The whole idea of the present and the previous chapter is, therefore, to be undogmatic and to construct a new

FIG. 18 Which one is the small black ball?

logic, which suits a new purpose. But this does not mean that anything goes. It seems, in fact, that we are starting to see some genuine drawbacks of fuzzy logic. Let me explain why, adapting examples from the British philosopher Dorothy Edgington and others. As we shall see, there is something more fundamental behind the collapse of the rules of excluded middle and non-contradiction, which will come back to haunt us even if we do not care much for logical purity.

Suppose we are looking at two rather similar balls, x and y, as well as some other balls which look very different (Fig. 18). Neither x nor y is particularly small; let's say they are small to degree 0.5. But while x is as black as can be, y is only borderline black (tending towards grey), at 0.5. The stated definition of 'and' predicts that '...is black and small' is true to the same degree for both balls, since the minimum of $v(\text{Black}(x))$ and $v(\text{Small}(x))$ is just as low, at 0.5, as the minimum of $v(\text{Black}(y))$ and $v(\text{Small}(y))$. Yet, if you ask me 'Bring me a small black ball', you would surely be happier if I bring you x rather than y. Taking only the *minimum* value into account makes all values that surpass the minimum irrelevant.

In response to problems of this kind, a popular alternative to the definition of conjunction based on minimum values is a definition that *multiplies* values. Although this is attractive in the example above, the new rule misfires if the two propositions are synonymous, as when [x is small] and [x is little] are conjoined: by multiplying the values of the two conjuncts with each other, one obtains a value lower than the orginal values (unless they happened to be completely true or completely false). The fact that a conjunction of two identical statements has a lower value than either of its conjuncts seems hard to defend.

These problems arise as a result of the way fuzzy logic understands 'and'. The situation with respect to 'or' ($\vee$) is similar. 'Or' is usually treated as the mirror image of 'and'. More precisely, the truth value of a statement of the form $p \vee q$ is equal to the *maximum* of the value assigned to p and the value assigned to q. Once again, this is often quite reasonable: if p and q are each either fully true (with a value of 1) or fully false (with a value of 0), for example, then this definition behaves classically: the value of 'the Pope lives in Rome or the Pope lives in Istanbul' comes out as a solid 1.

On the other hand, statements such as $\text{Short}(x) \vee \neg\text{Short}(x)$ do not always come out true: if x is sitting somewhere in the middle of the scale, then the value $\text{Short}(x)$ is 0.5, therefore the value of $\neg\text{Short}(x)$ is $1 - 0.5$, so the statement as a whole is assigned the value 0.5 once again instead of the usual value 1. In other words, the law of excluded middle stops being generally valid. Or consider a ball whose colour is halfway between red and pink: intuitively, the claim that this ball is 'red or pink' should have a very high truth value. (If someone said 'Please pick up the ball that's red or pink', that would be a perfectly clear reference to the ball.) Yet, this is not what taking the maximum predicts. In other words, using the maximum value to compute the value of a disjunction appears to have drawbacks similar to the ones associated with using the minimum value to compute the value of a conjunction. The issue at stake here is none other than that of Kit Fine's penumbral connections (Chapter 8). Penumbral connections will come back to haunt us when trying to apply fuzzy logic to sorites.

At the heart of these issues lies the fact that fuzzy logic, in logicians' parlance, is *truth functional*. This means that the truth value of a compound statement depends solely on (i.e. is a function of) the truth values of its parts. More precisely, the fact that $v(p \& q)$ is determined by nothing else than $v(p)$ and $v(q)$ makes it insensitive to the question whether p and q are related. The same is true for statements of the form $p \vee q$ and, as we shall later see, $p \rightarrow q$. In recognition of these limitations, we shall later look for degree theories that are not truth functional.

These concerns do not prevent fuzzy logic from juggling degrees of truth in a sophisticated and useful manner. To see how, the next section will take a brief look at some of the practical ways in which fuzzy logic has been applied.

DIALOGUE INTERMEZZO
ON FUZZY LOGIC

Science is a rational pursuit, but it is carried out by creatures of habit and social interaction. If you have been educated as a logician, you talk with logicians, read what logicians read, and so on; if you're an engineer then you hang out with different people and read different things. And if, on occasion, you encounter unusual ideas then chances are you won't like what you see. To the lay person, engineering and logic seem similar, but to the initiated, the differences seem enormous. Enter Lotfi and Kurt, two academics who run into each other at a campus coffee bar after an interdisciplinary research seminar on vagueness. Apparently, the famous Brazilian footballer Ronaldinho had featured in the discussions.

LOTFI: *It's a funny old debate, isn't it, about the sorites paradox? To hear highly paid academics debate stupid questions about people's heights, like 'Is Ronaldinho tall, yes or no?', is not what I call useful.*

KURT: *I'm not sure I find the debate stupid or useless. I'm puzzled by the facility with which people use vague expressions. Every 3-year-old uses them, yet no one seems to understands how they work. And surely the size of a footballer is of some interest. Heading the ball, you know . . .*

196

LOTFI: *I'm not saying the debate is stupid, but to force a decision between yes and no is stupid. In the case of Ronaldinho, the answer is somewhere in the middle. He's not really tall, and certainly not particularly short from what I hear. Suppose his height is about 180 cm, then he may be tall to a degree of, say 0.7, depending. To say that he's tall (or not tall, for that matter) is unhelpful.*

KURT: *You'd prefer to call him 70 per cent tall, in other words?*

LOTFI: *Exactly...*

KURT: *Exactly?*

LOTFI: *Well, in a manner of speaking. Whether 180 cm counts as 0.6, 0.8, or something in between does not keep me awake at night. I'm assuming that one uses some reasonable method for deciding on the right number.*

KURT: *And the method...*

LOTFI: *...little hinges on it. As long as it's consistently applied. For example, you could set the minimum height for all footballers alive, let's say, as short to degree 1. The maximum gets degree 0. You could then, for example, connect the two points with a straight line. That's one way. If you prefer another method, fine. You can also take various aspects of the context into account, for example that he's Brazilian, or his role on the football pitch.*

KURT: *I can see certain benefits in such an approach. Every height would correspond with a different degree of shortness, and this would allow you to say that if x and y are similar in height then the truth values of 'x is short' and 'y is short' are similiar though possibly different.*

LOTFI: *That's right. I bet you can see how I use this to explain sorites.*

KURT: *You call that an explanation? Remember that you've got to do the business regardless how close together x and y are. Even a difference of a millionth of a millimetre is enough to justify different truth values, or else sorites will punish you.*

LOTFI: *Yeah—so I give them all different values. I use the whole continuum between 0 and 1, so there are always enough numbers to choose from. So what?*

KURT: Well, it means that these figures of yours—I mean truth values like 0.5 and 0.51 and 0.501 and whatnot—cannot reflect what people perceive. If x and y are really similar in height, so similar that no one could possibly see the difference, then you're still going to have to assign different truth values to 'x is short' and 'y is short'.

LOTFI: Yes, that's the price we pay … Other accounts are forced to assign the values 0 and 1 to such sentences! They have to locate the boundary somewhere, don't they? Surely that's worse. By using an infinite number of truth values, I minimize the damage. What's wrong with multiple truth values?

KURT: Well … you're giving up the beauty of classical logic, for one thing. But I guess that's not an argument that would appeal to you. Let me put it differently: I don't know what all those truth values of yours mean! You'd never be able to verify them in an experiment with real people, surely. Suppose you show a subject these two people x and y that we were talking about; do you really believe your subject is going to tell you that x and y are short to different degrees? Let alone the specific degrees that you just suggested!

LOTFI: Maybe not. But suppose I let my subject make the same judgement again and again, maybe using a slider that can be operated by hand, shifting it to the left for the truth value 0 and to the right for the truth value 1, and everything in between, with every position corresponding to a number in between. (I think this is called magnitude estimation.) If I took the average of a large number of my subject's attempts then it seems plausible to me that these averages would be different for x and y. You see? If y is slightly larger than x, then in the long run y must be assessed as being less short.

KURT: Nice try! And plausible enough if x and y are a few millimetres apart. But not if the difference is, say, a billionth of a millimetre! Or one billionth of one billionth if you prefer. Remember that sorites does not hinge on the size of the difference: it can be as small as you want. Every measurement tool has intrinsic limitations, beyond which it fails to discriminate. And if such small differences in height cannot exist then we swap to a different domain where they can exist.

LOTFI: How do you mean?

KURT: *Temperature is as good an example as any. Suppose you stick a digital thermometer in your cappucino. At certain intervals, the thermometer will register a change. Just now, it might indicate 60 degrees Celsius; in ten seconds or so it may shift to 59.9. Do you really think the coffee does not change temperature until the thermometer shifts to 59.9? Of course it does! It changes all the time. But your thermometer will register the change only once the change has become large enough. I don't need to tell you that things are the same if your thermometer uses more digits. (And yes, the thing might occasionally flipflop between one position and the next, but that won't help your argument.) Your thermometer may be good, but it won't be perfect. As far as it is concerned, there really is no difference between certain temperatures. No measurement instrument is perfect, and certainly not our eyes, ears, and the rest of it. You might even run into a situation where your subjects get things the wrong way around, calling one person taller than another when it should really be the other way round.*

LOTFI: *You're splitting hairs. On reflection, I disagree when you say that I have to be able to assign different values to all different heights. If my subject occasionally assigns the same values to the heights of x and y, where in reality x and y differ slightly, then fine! If we measure increasingly larger people, then soon enough we'll reach a point where a different value will be assigned. If this happens when measuring z, for example, then the degree of shortness will be the same for x and y, but that of z will be smaller. Differences like the one between y and z will allow me to prevent sorites. And of course, my numbers are only an approximation. To ask more is . . . is a misunderstanding. It's a misunderstanding of what science is about. You want truth, eh? Forget it, Kurt. The best you're going to get outside your pure mathematics is approximations. Newtonian physics? An approximation. Quantum physics? A better approximation. To ask for more is naive!*

KURT: *My point was not that your truth values are incorrect: they're worse than incorrect, if you get my meaning. They are messy. Arbitrary. But why am I talking with an engineer anyway? Time to get back to the office, prove some theorems.*

Fuzzy logic and the sorites paradox

We have seen that the sorites paradox hinges on premisses of the form

$$\text{Short}(x) \rightarrow \text{Short}(x + 1)$$

where the individual $x + 1$ is always slightly taller than x. Each of these premisses takes the form of a conditional. We therefore need to ask how conditionals in general are treated in fuzzy logic: we have dealt only with negation, conjunction, and disjunction so far. Once again, several possible treatments have been proposed. One option that comes naturally to a logician (see the section on classical logic in the previous chapter) is to read a statement in the form 'If p then q' as equivalent to '(not p) or q'. This is a reasonable move if you think about such sentences as 'If there is free drink at the party then John will be present', which comes out as 'There is no free drink at the party or John will be present'. In the sorites conditionals, however, this produces erratic results, assigning very different values to the different premisses even though, intuitively, they are all alike. Worst of all, the values in the middle of the crucial interval receive values almost as low as 0.5. Suppose, for example, that Mr 170's degree of shortness is 0.50 while that of Mr 171 is 0.49. Then the value of the conditional is computed thus:

$$v(\text{Short}(170) \rightarrow \text{Short}(171)) =$$
$$v((\neg\text{Short}(170)) \vee \text{Short}(171)) =$$
$$\text{maximum}(v(\neg\text{Short}(170)), v(\text{Short}(171))) =$$
$$\text{maximum}(0.50, 0.49) = 0.50$$

The reason why this truth value comes out so low is that the formula for computing it does not take into account that the two halves of this particular conditional are strongly connected with each other, given that they talk about individuals of such similar heights. Several repairs have been proposed. The one that has become more or less standard focuses

directly on the difference between the truth values of Short(*x*) and Short
(*x* + 1) and lets the truth value of the conditional be determined by this
difference. If *x* is slightly shorter than *x* + 1, then the truth value of the
conditional is close to 1:

If $v(p) > v(q)$ then $v(p \rightarrow q) = 1 - (v(p) - v(q))$
Otherwise $v(p \rightarrow q) = 1$

Applying this to the sorites conditionals, we get

If $v(\text{Short}(x)) > v(\text{Short}(x + 1))$ then
$v(\text{Short}(x) \rightarrow \text{Short}(x + 1)) =$
$1 - (v(\text{Short}(x)) - v(\text{Short}(x + 1)))$

Given that $v(\text{Short}(x)) - v(\text{Short}(x + 1))$ is always very low, this makes
each of the sorites conditionals almost perfectly true. Yet, the ultra-short
individual with which a typical sorites series starts will give statements of
the form Short(*x*) a truth value close or equal to 1, while the giant at the
end will give these statements a truth value close or equal to 0, which is
of course as it should be. To see how this works, consider the first
conditional:

Short(150) $\rightarrow$ Short(151)

Suppose $v(\text{Short}(150)) = 1$. We know that $v(\text{Short}(150) \rightarrow \text{Short}(151)) =$
$1 - (v(\text{Short}(150)) - v(\text{Short}(151))) = 0.9$. Given that $v(\text{Short}(150)) = 1$, it
follows that $v(\text{Short}(151)) = 0.9$. In other words, we can use the conditional to say something about the shortness of the second individual in
the sorites series (i.e. the one we named 151), but something is lost in the
process, since the argument will support the shortness of Mr 151 only to a
degree of 0.9. The shortness of the next individual can be supported only
to a degree 0.8, and so on. Once again, all of this seems very much on the
right track.

A small wrinkle in this story is the fact that the standard definition of
conjunction, which uses the minimum of the values of the conjuncts,

will give an uncomfortably high value to the conjunction of all the different conditionals of which the sorites paradox is made up: in our example, the conjunction ((Short(150) → Short(151)) and (Short(151) → Short(152)) will have the minimum of the truth values 0.9 (for the first conjunct) and 0.9 (for the second conjunct), which is still an impressive 0.9. Adding further conjuncts of the same kind does not change the story. The usual response is to move to the multiplicative definition of conjunction, whereby $v(p \, \& \, q) = v(p) \times v(q)$. In this way, one explains that each of the conditionals is nearly true when taken individually, but taken together in sufficient numbers, their conjunction is nearly false, a bit like in Kamp's context-based approach, which was discussed in the previous chapter.

How should one assess the analysis offered by fuzzy logic? On the one hand, there is no denying the cogency of fuzzy logic's basic insight, that vague statements can be accurate to different degrees. And, to be able to explain away a paradox by arguing that all the steps in the argument are *nearly* correct is as good as it gets. It is, for example, neatly in line with Steven Pinker's characterization that 'the slop accumulates' with each step in the argument. What is more, this perspective on the paradox makes it understandable why we find its reasoning plausible, for to confuse absolute and approximate truth is surely forgivable. Which of us would claim always to know the difference between what's perfect and what's very nearly perfect?

There is, however, a more problematic side to fuzzy logic. We have seen that the truth functionality of this approach makes it difficult for fuzzy logic to deal with penumbral connections: this limitation makes fuzzy logic unable, for example, to distinguish between an arbitrary conjunction $p \, \& \, q$, on the one hand, and a very special conjunction such as $p \, \& \, \neg p$ on the other, where one part negates the other. The problem is that if q and $\neg p$ have the same truth value, fuzzy logic is unable to distinguish between them. Although this is acceptable from a purely mathematical point of view, truth functionality does limit fuzzy logic's appeal if it is to be used as a source of insight into logical argumentation, as was argued by Kit Fine, for example, when he discussed *penumbral connections* (see Chapter 8).

The place where this hurts fuzzy logic's treatment of sorites most is in the conditional. We have seen that, in order to avoid strange truth values, a tailor-made interpretation had to be attached to the sorites conditionals. This formula, however, would be implausible in some other cases. To see this, suppose once again that the statement Short (170)—right in the middle of the domain—received the truth value 0.5, so it is thoroughly unclear whether this individual is short or not. Under this assumption, it seems reasonable to assign some value close to 0.5 to the statement Tall(170) as well. Now strange things start to happen. For if the value of Tall(170) comes out as 0.5 or higher then the formula above tells us to consider Short(170) $\rightarrow$ Tall(170) as entirely, impeccably true. If the value of Tall(170) comes out slightly below 0.5, say at 0.5 minus some small number ϵ, then the conditional comes out as $1 - \epsilon$, which is almost perfectly true. This seems just wrong. Surely, we do not want to say that if a person of 170 cm is short then this person is also tall.

Probabilistic versions of many-valued logic

We have seen that degree theories bring an appealing subtlety to questions about vagueness. This appeal is further enhanced by the fact that degree theories can be applied to all versions of the paradox, whether they involve indistinguishables or not: for unlike some context-based solution for example, the many-valued approach does not rest on an analysis of indistinguishability. But we have also learned that fuzzy logic's approach to propositional connectives is, from our point of view, undesirable, because it fails to take semantic connections between sentences into account. Might it be possible to keep what is good about fuzzy logic while repairing its flaws? Can we, in other words, have our cake and eat it?

Looking over the battlefield, what we seem to need is an account of vague predicates that combines the many truth values of fuzzy logic with supervaluationism's ability to explain penumbral connections. As it happens, the kernel of an account with these properties was sketched long ago, by Max Black in 1937. The core of this approach is to think in terms of the probability of someone agreeing with a particular statement. Recently, Dorothy Edgington (1992, 1996) has defended a similarly probabilistic approach. Edgington used the word *verity* when talking about her probabilistically inspired degrees of truth. Consider once again a Fine-style disjunctive statement such as 'This blob is red or this blob is pink' (or 'You are tall or of average height', for example). What is the probability of someone agreeing with this disjunctive statement? Certainly not the maximum of the probabilities of its two disjuncts, as fuzzy logic would have it. Something like the added probability of the first and the second disjunct seems closer to the mark. But, generally speaking, addition is only a first approximation to a logic for disjunction. To see why, let us first talk about 'real' probability, unrelated to vagueness.

Consider the probability of 'I throw an even number *or* a number divisible by 3' with a die. The probability of this disjunction does not simply depend on the probability of my throwing an even number (which is ½) and the probability of throwing a number divisible by 3 (which is ⅓). Because the number 6 is both even and divisible by 3, the two statements overlap, so if we simply added the two probabilities, the overlap would be counted twice! By subtracting the probability that a number is thrown which is both even and divisible by 3, standard probability avoids double-counting:

$$p(A \lor B) = (p(A) + p(B)) - p(A \ \& \ B)$$

Now back to vagueness: along similar lines, one can argue that the probability of someone agreeing with the disjunctive statement 'This blob is red or this blob is pink' should equal the probability of someone agreeing with the first disjunct *plus* the probability of someone agreeing with the second disjunct, *minus* the probability of someone agreeing with both. If Fine is right about colours then the subtracted probability is nil in this particular case, but this may be different for 'This blob is brown or this blob is dark', because it is in the nature of the penumbral connections between brown and dark that they overlap. We shall see the importance of this point more clearly when the probabilistic approach is applied to the sorites paradox. What counts here is that a probabilistic perspective on red and pink allows us to honour the connections between the two concepts.

A remaining weakness of these ideas lies, I believe, in a lack of clarity concerning the kind of probability involved. To put it simply, it is not always clear what it means to say that the truth degree of a given statement is n. In what follows, I shall offer two different interpretations: an abstract one, and a human-oriented one, which I believe to be the more interesting of the two. We start with the former, since it is easiest to explain.

Probabilistic logic: The abstract version

Cast your mind back to the view of vagueness expounded by supervaluations, focusing on a person's height as before. We asked about all the different ways in which the concept 'short' can be defined. The standard answer could be pictured by imagining a line on the wall, from the floor to the ceiling, where every point defines a threshold for shortness.

Now suppose I took a dart and hurled it towards the wall. For the sake of argument, we assume that any height between 150 cm and 175 cm is equally likely, while the dart will never land outside that range. The point where the dart hits the wall defines a threshold for shortness, in the sense that everyone below that point will count as short and everyone else as not short. What is the probability that my dart ends up judging a person of 162.50 cm short? About 0.5. What's the probability of calling a person of 163 cm short? Slightly lower. The taller the person, the lower the chance that my dart will end up above this person, causing him to be

205

counted as short. The probability of a statement p being true is equated with the probability that a throw of the dart will make p true. Our final move is to equate the degree of truth of a proposition with its probability (or verity) in the sense just explained.

This story has many of the properties that our discussion has shown to be desirable. All the usual laws of probability apply, for example. Using v for verity, and $\times$ for multiplication, we have:

(a) $v(\neg\phi) = 1 - v(\phi)$

(b) $v(\phi \vee \psi) = (v(\phi) + v(\psi)) - v(\phi \,\&\, \psi)$

(c) $v(\phi \rightarrow \psi) = v(\psi \mid \phi)$

(d) $v(\phi \,\&\, \psi) = v(\phi) \times v(\psi \mid \phi)$

Definition (a) is unproblematic, since the probability of being below the threshold must equal the complement of the probability of being at or above the threshold. Definition (b) is in accordance with what was observed regarding 'I throw an even number or a number divisible by 3', which is why (b) subtracts $v(\phi \,\&\, \psi)$ from $(v(\phi) + v(\psi))$. Definition (c) hinges on the notion of a *conditional probability*: $(\psi \mid \phi)$ is the probability of ψ given ϕ, the probability that someone who knew that ϕ is the case should assign to the probability of ψ. For example, the probability that a natural number is divisible by 4 *given* that it is even equals 0.5; under the same condition, the probability that such a number is divisible by 8 is 0.25, and so on. Definition (d), which makes use of conditional probabilities as well, merits more explanation.

A novice to probability, when asked about the probability of two events, ϕ and ψ, jointly taking place might say $v(\phi) \times v(\psi)$. After all, the probability of throwing a 3 with a fair die is 1/6, the probability of throwing a 4 is also 1/6, therefore the probability of throwing first a 3 then a 4 is 1/36. This, however, fails to take into account that there may be connections between the two events. What if we have no guarantee, for example, that the die is fair? In that case, the first throw of 3 might make 3 likelier than other outcomes. Or, closer to sorites, what if we want to calculate the probability that 163 cm *and* 162.5 cm will be counted as short? We are not dealing with independent events here, for if a throw of the dart causes 163 cm to count as short then automatically 162.5 will also count as short: multiplying the two probabilities would hugely underestimate the likelihood of the joint occurrence of the two events. This is why (d) is formulated using the *conditional* probability of ψ given that ϕ.

It's worth digressing here for a moment. For in daily life, a failure to acknowledge that different events may be connected with each other can have grave consequences. A few years ago, for example, a number of people in Britain, who had seen several of their babies die from cot death, were wrongfully convicted for killing their babies. This came about as a result of expert testimony by a distinguished paediatrician, based on the dictum, 'one sudden infant death is a tragedy, two is suspicious and three is murder, until proven otherwise'. The statistical principle behind his dictum was simple yet dead wrong: that the probability of a combination of events can be computed by simply multiplying the probabilities of the individual events with each other. The paediatrician argued that since the probability of one cot death is about 1 in 8.54 million, the probability of two cot deaths must be about 1 in 8.54 × 8.54 million, amounting to 1 in 73 million. Incredibly, this expert had failed to realize that the two events could be linked in some way, for example by a common genetic cause, so the probability of the second death *given* the first is much higher than 1 in 8.54 million. Needless to say, the consequences of this error in probability logic (which went unnoticed in court) must have been devastating for all involved.[3]

What do all these definitions mean for the sorites paradox? We have seen the implications of the dart-throwing analogy for the degree of truth of simple statements of the form 'so-and-so is short': the perspective is holistic in the same way as the supervaluational perspective. We look at *all* possible thresholds, identifying the truth of the proposition with the likelihood that a throw of the dart will cause someone to be counted as short. Presumably, this degree is extremely high for the first individual in the sorites series, Mr 150, say $v(\text{Short}(150)) = 1$. How about the crucial premisses? Consider, for example

$$\text{Short}(150) \rightarrow \text{Short}(151)$$

The value of this conditional must equal $v(\text{Short}(151) \mid \text{Short}(150))$: the probability that 151 counts as short assuming that 150 counts as short. A good way to understand this is in terms of lengths of line segments. Suppose the darts can be thrown between 150 cm and 175 cm, a length of 25 cm. The only part of the line where, if the dart were to land there, 150 would be called short while 151 would *not* be called short is the segment between 150 cm and 151 cm. It would be reasonable to represent the probability that 150 is short but 151 is not, as the length of this 1 cm line segment divided by 25 cm, yielding a low value of 0.04. This probability represents the case where the conditional is false; it is *true* over a stretch of 24 cm of the total 25 cm, giving it a truth value of about 0.96.

This analysis suggests the following pattern of explanation. First, each of the sorites conditionals has a high value, which nevertheless falls short of full truth. Each step of the sorites argument applies a premiss of this kind to a proposition that 'so-and-so is short', where this proposition itself may either be fully true (in the very first step of the argument) or less so. What happens when *uncertain* premisses enter a logical argument? Symbolic logic was not designed to cater for situations of this kind, but recent investigations have clarified some important issues. Dorothy Edgington, building on the work of Ernest Adams and others, argues for the following principle, in which the *uncertainty* of a proposition is understood as 1 minus the probability of that proposition.

An argument is valid if and only if it is impossible for the uncertainty of the conclusion to exceed the sum of uncertainties of the premisses.

208

For example, suppose you believe A and B each has a probability of 0.99, which corresponds with an uncertainty of 0.01. You are allowed to use these propositions to draw inferences, but this will not allow you to attach a probability of more than 0.98 to your conclusions, whose uncertainty of 0.02 is jointly inherited from the two premises.

Let us apply this idea to sorites. First, we combine the premiss Short (150), whose truth degree is 1, with the conditional Short(150) → Short (151), whose truth degree is 0.96. We derive the conclusion, Short(151), with truth degree 0.96, as we have seen. The second step allows us to derive the conclusion Short(152), but only with a truth degree diminished by another 0.04, totalling 0.92, and so on. The halfway mark of 0.5 is reached around the middle of the 150–175 cm range, and once we arrive at the end of the range, we derive $v(Short(175)) = 0$. Each subsequent statement is derived with less confidence: a confidence of 0 in a statement is tantamount to a confidence of 1 in its negation: Mr 175 is not short, in other words. To put the icing on the cake, this account gives the conjunction of all premises a value of 0, even though each premiss itself has a value of fully 0.96. No ad-hoc measures need be taken, everything falls out automatically.

I do have one worry, and it involves the tolerance principle. For, presumably, the darts in our story were sharp enough to separate any two heights. They were not real darts, in other words, but mathematicians' darts without any width—just like a mathematician's line has no width. But isn't this exactly what we have been attempting to avoid? We have replaced the one crisp boundary between, say, short and not short, by infinitely many crisp boundaries. One such boundary lies, for example, between all people who are at least 170 cm tall and all those who are shorter. This, one might argue, mistakenly ascribes unlimited powers of perception to us in just the same way as classical logic. In Mark Sainsbury's (1990) words: you do not improve a bad idea by iterating it infinitely many times.

An engineer might respond that fuzzy truth values are only approximations, and that a good approximation is better than a bad one. When your digital watch says 11:56:32, you do not expect this measurement to be *correct*, since the actual time may be many seconds earlier or later: yet the figure displayed by your watch could be more useful to you than to hear that it is noon. This response has considerable force. Yet, I find it unsatisfactory, in a theory that aims to shed light on the sorites paradox, if the theory does not come to terms with the fact that people's perceptual judgements are different and variable. It was, after all, this fact which caused problems in the previous chapter, and which led us to explore an approach that took variations in people's perception judgements into account. Therefore, let's try a less abstract take on these matters, which takes people's judgements as its point of departure.

Probabilistic logic: The human-oriented version

The roots of a human-oriented approach to probabilistic logic can be found in the writings of that oldest of degree theorists, Max Black. Black defined the truth value of a vague statement as the proportion of competent speakers who would be prepared to assent to that statement in a situation in which they are aware of all the relevant facts. To make this idea more precise, he sketched a thought experiment along the following lines.

Participants are asked, of a given person x in full view of them, whether they consider that person short or not. Each person is shown in splendid isolation, so participants have no easy way to compare the people whose heights they are asked to judge. Also, they are shown in such a way that only their height is visible, and they cannot be identified. Participants are asked to choose between yes and no: there are no intermediate options, and no 'don't know' option either. If 75 out of a total of 100 people give an affirmative answer then the truth degree of the statement 'x is short' is 75/100, or 0.75. Suppose you are one of the participants. You may not know other participants' answers to the question, all you know is that *you* said yes. Accordingly, you might not know the truth *degree* of the statement, since this can be computed only from all answers taken together.

210

Crucially, we cannot assume you to be consistent in your judgements. If the same person x appears in front of you on ten separate occasions, you may or may not judge x in the same way every single time, particularly if x is a borderline case. (Recall that you are unable to identify x and say 'We've met before!') Similarly, if you are shown *another* person y, whose height is very close to that of x, you may or you may not judge y in the same way as x. Like the agents of the previous chapter, your judgements may be fickle, depending on the condition of your eyes and brain. If we can make this empirical perspective work then we will have dodged the tolerance principle very handsomely while still honouring the ideas behind it. For suppose x and y are shown to you 1,000 times. Now because x and y are so similar in height, it stands to reason that they will give rise to *similar* sequences of judgements: maybe x will be judged to be short 342 of the 1,000 times and y 339 times. It would be little short of a miracle, however, if x and y were judged to be short the *exact* same number of times.

How, under this interpretation, should compound statements be judged? Consider the sorites conditionals. What is a sensible probabilistic account that might be matched to Black's experiment? We want to keep the usual laws of probability, of course, but how can we interpret *conditional* probability? How, in other words, should one interpret a statement such as $v(\text{Short}(161) \mid \text{Short}(160)) = n$? If all the judgements of all the participants in the experiment are lumped together, it is difficult to see how a judgement on the truth or falsity of Short(160) can influence the truth degree of Short(161). But if the judgements of each participant are grouped then a natural account suggests itself. Suppose the heights of a number of people are judged by a given subject. For simplicity, let each subject judge a given person's height only once. This gives rise to a *judgement set*: a set of judgements, by one subject, about the heights of a sequence of people.

In the simplest case, each judgement set could be characterized by one threshold, where 'short' judgements give way to 'not short' judgements. One subject might, for example, consider everyone at or below 160 cm to be short while considering everyone else not short. If all judgement sets were of this kind, the situation would bear a strong resemblance to the dart-throwing scenario discussed before. But subjects' behaviour might be a good deal less systematic, and consequently judgement sets may not be *admissible* in the sense in which this concept was used in the previous chapter. To focus the mind, here are a few possible judgement sets. $S(149)$ abbreviates the judgement that Mr 149 is short.

Rosanna's judgements:

$\{S(149), \ldots, S(153), \neg S(154), S(155), \neg S(156), \neg S(157), \neg S(158), \ldots, \neg S(175)\}$

Roy's judgements:

$\{S(149), \ldots, S(153), S(154), S(155), \neg S(156), \neg S(157), \neg S(158), \ldots, \neg S(175)\}$

Hans's judgements:

$\{S(149), \ldots, S(153), S(154), S(155), \neg S(156), S(157), \neg S(158), \ldots, \neg S(175)\}$

Joe's judgements:

$\{S(149), \ldots, S(153), S(154), \neg S(155), \neg S(156), S(157), \neg S(158), \ldots, \neg S(175)\}$

Tim's judgements:

$\{S(149), \ldots, S(153), S(154), S(155), S(156), S(157), \neg S(158), \ldots, \neg S(175)\}$

Roy and Tim adhere to the principle of admissibility, since they change their minds only once. Roy changes his mind at the height of 156 cm. Tim, who might be used to slightly taller people, changes his mind at 158 cm. The other three participants change their minds several times. Joe, for example, called the person of 154 cm short, the one of 156 cm not short, and the one of 157 cm short, before settling on a regular pattern.

Now consider the statement Short(150) $\rightarrow$ Short(151), whose verity equals the conditional probability of Short(151), given Short(150). For any statement p, we understand $v(p)$ as the probability that an arbitrarily chosen participant would judge p to be true. This makes $v(\text{Short}(150)) = 1$, because each of our five participants called Mr 150 short. The same is true for Mr 151, giving our conditional a value of 1. A more interesting example is Short(153) $\rightarrow$ Short(154). Once again, the antecedent, Short(153), is true according to 5/5 subjects, hence the probability of picking a participant who called this statement true is 1. The probability of picking a subject who called the *consequent* true is 4/5; therefore, the conditional probablity $v(\text{Short}(154) \mid \text{Short}(153))$ is also 4/5. The next conditional, Short(154) → Short(155), shows the relevance of conditional probability: it has a probability of 3/4, because once we know that someone calls Mr 154 short (which is true for Roy, Hans, Joe, and Tim), the chance that he or she calls Mr 155 short is 3 (Roy, Hans, Tim).

A human-oriented account along these lines has much in common with its more abstract predecessor: it asserts that most of the conditionals relevant for the sorites paradox have high truth values, while this is not true for their conjunction. Also, like its predecessor, it can use Edgington's analysis of logical reasoning to explain that very little credence can be attached to the conclusion of the sorites argument. The difference is that the human-oriented account acknowledges that people can vary in their judgements, particularly where borderline cases are concerned.

A comparison with the epistemicist approach might be useful. Like epistemicism, our proposal assigns definite truth values to expressions of the form 'x is short'. But unlike epistemicism, it is the *gradedness* of these values that helps us find a plausible version of the tolerance principle, and which allows us to explain why sorites is plausible yet fatally flawed. Furthermore, like epistemicism, the present approach denies that speakers *know* the truth values of all the statements in question. On the present theory, this is because speakers differ and because even one speaker's judgements might be different at different times. There is, in other words, nothing to know, except how self and others use the word 'short'. There is no fixed meaning of which speakers are ignorant, let alone a crisp one.

What's wrong with degrees?

Truth is a hazardous topic to mess with. In the European Middle Ages, some theologians were starting to find it difficult to reconcile the teachings of the Bible with those of such ancient philosophers as Plato and Aristotle, which were held in tremendous esteem. In response, these theologians proposed a theory of 'double truth', which asserted that a statement could be true in science yet false in religion, or the other way round. According to them, science and religion each had their own realm of influence, as it were, which could never be in conflict with the other. This would have given science a free rein, of course, to pursue its own goals without having to worry about papal condemnation. The Church, however, was quick to see the threat and condemned these poor souls, threatening them with excommunication. Degree theories are logically different from the theory of double truth, and their proponents are not excommunicated. It is, however, conceivable that modern heretics find it difficult to get a good academic position. Times have changed, but it is never risk-free to advocate an unpopular view.

It is striking, for example, that many discussions of the sorites paradox mention fuzzy logic as the only degree theory, only to dismiss it in a hurry. Rohit Parikh is a good example. His otherwise excellent 1994 paper devotes just one page to fuzzy logic, stating without further explanation that 'Since our interests are in clarification rather than in the design of cameras (...) we will proceed to the next approach', alluding to fuzzy logic's practical applications. Off-hand dismissals of this kind are surprisingly common. At this point it is tempting to recall the arguments of Dawkins (with his tyranny of the discontinuous mind) and Blastland and Dilnot (with their 'false clarity': see Chapter 1), who argued that people like to view reality in terms of predicates that are always either totally true or totally false. We have seen that this might explain our remarkable credulity when confronted with claims that 'the gene for' another disease has been found, that a substance has been found to be poisonous, or that x per cent of people in a given city are black, to repeat just a few of the examples discussed in the Introduction. Perhaps that theoreticians are suspicious of degree theories partly for irrational reasons.

214

Other reasons come to mind, such as a justified awareness that classical logic is a well-understood mathematical system of tremendous elegance and power. Academic disciplines such as engineering, whose mathematical backbone lies in calculus rather than in symbolic logic, are not affected by false clarity in the same way as linguistics and philosophy. But the position of classical logic alone cannot explain the resistance against degrees, given that other non-classical logics have taken hold, in harmonious collaboration with classical logic. The reasons, therefore, must have something to do with the *particular* way in which degree theories deviate from classical logic.

One factor that may have hampered the acceptance of degree theories is that their best-known representative, fuzzy logic, has certain weaknesses that have caused other degree theories to become tarred with the same brush. We have seen that an adherence to truth functionality proved to hamper fuzzy logic's ability to model logical arguments such as sorites. There is irony in this, because fuzzy logicians are often depicted as iconoclasts, yet it can be argued that they fail because they have not been radical enough. Truth functionality is fine in classical logic, but when multiple truth values are allowed, it can stand in the way of progress.

It has been suggested that these weaknesses are relevant only for theoretical pursuits, and that they are harmless in the practical applications for which fuzzy logic is best known. I am not convinced that this is correct. Every truth-functional way of defining logical operators runs into difficulties in some situations. Fuzzy logic's way around this is to use whatever definition gives the best results in a particular application. The truth degree of a conjunction, for example, is sometimes computed by taking the minimum value of the conjuncts, and sometimes by multiplying these values, when this is felt to give better results. But the fact that fuzzy logic has no *theory* about this means there is a risk of choosing a conjunction that is wrong for one's application. Fuzzy logic is not a full-blown logical theory, in other words, but a do-it-yourself pack with which you can define your own theory. Basically, if you use fuzzy logic, you are on your own. We shall expand on this point in the following chapter, where applications of fuzzy logic are discussed.

Furthermore, a procedure of this kind does not *explain* much. When a theoretician asks for an explanation of the sorites paradox, for example, she wants a consistent perspective on what sentences of a particular form mean, and what inferences would be justified. Invoking a particular type of conjunction just because it gives the totality of premises an appropriately low value doesn't quite cut the mustard. To make things worse, fuzzy logicians have sometimes presented their framework *as if* it were a full-blown theory. As a result, fuzzy logic may have diminished the reputation of degree theories in general.

Some people, such as the philosopher Rosanna Keefe in her book *Theories of Vagueness*, have gone beyond the ritual dismissals and attempted to argue systematically against degree theories. Perhaps the most powerful weapon in their armory is the existence of multidimensional concepts, which juggle *combinations* of gradable dimensions. If John has a quicker wit than Joe, while Joe is better at solving analytical puzzles, who is the more intelligent of the two (see. Chapter 3)? Degree theories force us to say that the predicate 'intelligent' applies either more to Joe, or more to John, or to exactly the same degree in both cases. Similarly, a degree theory forces us to assign a particular degree of 'goodness' to a car, even though the quality of a car depends on multiple features. In some people's opinion, this is unacceptable.

Admittedly, the logic of multidimensionality is tricky, for example because it can easily give rise to cycles. We saw this in the dialogue fragment in Chapter 3, where several reasonable decision procedures forced us to conclude that, of three candidates for an academic job, the first was better than the second, the second better than the third, yet the third was found to be better than the first. By the same token, we can easily be tricked into saying that we prefer one car over a second car, the second over a third, but the third over the first. I cannot see how this amounts to an argument against degree theories, however. It is, to begin with, not obvious why the theories of the previous chapter are not affected by the same problem, given that these theories will likewise have to determine whether it makes sense to discriminate between two of the above-mentioned people (or cars). Moreover, one could stubbornly insist that comparisons are *always* possible. In real life, we are able to compare cars, after all. If a decision procedure produces counter-intuitive results, then so much the worse for the procedure: we'll need to find another one.

After these theoretical chapters on the logic of vague expressions, let's now turn to more practical concerns by focusing, more explicitly than before, on computer simulations of human communication. It will be interesting to see what role the theoretical issues discussed in the last few chapters—revolving around such notions as context, degree, and indistinguishability—will play in a more computational and application-oriented setting.

Things to remember

◆ We have argued that approaches that rely on classical and partial logic are less well placed than degree theories to explain the utility of vague expressions in certain situations. The claim is not uncontroversial.

◆ Because degree theories have many truth values at their disposal, they are able to say that something is *almost* true. This gives degree theories an important advantage in the quest for a convincing analysis of the sorites paradox, because it allows them to say that each sorites conditional is almost true, but their conjunction is almost entirely false, and so is the conclusion of the sorites argument.

◆ Some of the resistance against degree theories appears to derive from our ingrained tendency to think in terms of all or nothing: what Dawkins called the tyranny of the discontinuous mind, and what Blastland and Dilnot called false clarity. Another contributing factor may be that fuzzy logic's limitations are sometimes seen as something absolute, rather than as a problem that may be overcome.

◆ Fuzzy logic has certain inbuilt features that hamper its practical applications and its relevance for the analysis of natural language, including paradoxes such as sorites. These limitations derive from the fact that fuzzy logic defines logical operators in a *truth-functional* way.

◆ Probability-based approaches along the lines proposed by Kamp and Edgington are degree theories that hold considerable promise, despite open questions about some of their properties, for example regarding their treatment of conditional statements.

◆ Degree theories abandon the strict dichotomy between vague and crisp concepts, explaining to what *extent* a concept is vague. The ultimate consequence of this position is that vagueness itself becomes a vague concept, because it has borderline cases: concepts whose vagueness is doubtful.

Part III

Working Models of Vagueness

10

Artificial Intelligence

When I was 20 or so, I paid a visit to an older acquaintance who had just acquired a new boyfriend. At one point during a long walk, out of earshot from her partner, my acquaintance confided in me. 'He works in artificial intelligence', she said circumspectly. I had not heard this expression before, but her grave delivery left me in no doubt about the weight of his activities. To hear that this boyfriend had managed to secure gainful employment was reassuring... but artificial intelligence? I thought it wise not to inquire further.

This chapter will introduce the goals that researchers in artificial intelligence (AI) have *actually* set themselves, and how their efforts at reaching these goals have come along during AI's first half century.[1] In the process, we shall get acquainted with some of the ways in which AI has dealt with vagueness.

A brief history of AI

There does exist a kind of artificial intelligence (AI) that is as ambitious as the term suggests. In its most pure form, the ambition of AI is to build working models of the entire human mind, and sometimes the human body too. Mary Shelley's novel *Frankenstein*, published in 1818, comes to mind, in which a student creates a monster from parts of corpses. Even earlier, in the seventeenth century, the Jewish myth of the Golem involved a similar feat, in which a Rabbi Loew of Prague builds a manlike creature from sticky river mud, to help his community who are going through difficult times. Understandably, he makes the Golem a bit larger and stronger than ordinary people—something that became an enduring theme of the robotic genre, to be played out in numerous novels and films.

It took AI until the 1950s to enter mainstream science. The year that is usually cited is 1956. Computers were only beginning to come out of the research laboratories: reportedly, by 1953, IBM had sold no more than nineteen computers. Computer memory and processing power were extremely limited. Even the largest computers in existence could not store more than about 5 million bytes; it would be years before floppy disks and PCs were to appear on the scene. Most important of all, people's ability to construct useful computer programs was still in its infancy. Yet, computer scientists were starting to dream. A brainstorming event known as 'the Dartmouth conference' was organized around the following proposal:

> We propose that a (...) study of artificial intelligence be carried out (...). The study is to proceed on the basis of the conjecture that every aspect of learning or any feature of intelligence can in principle be so precisely described that a machine can be made to simulate it. An attempt will be made to find how to make machines use language, form abstractions and concepts, solve kinds of problems now reserved for humans, and improve themselves.

Two years later, the researchers Simon and Newell predicted that significant advances in a number of major areas of AI would be achieved in just ten years. Among other things, they predicted that by that time computers would have learned to play chess better than any human. Fifty years later, computers have become much more 'intelligent' in many respects, including the areas on which Simon and Newell concentrated. At the time of writing, for example, the world has started to accept that some computer chess programs—answering to the names of 'Deep Blue' and 'Fritz'—are stronger than even the best human players. Simon and Newell did eventually have their way, though it happened thirty years later than they predicted.

With hindsight, it seems hard to understand the unbridled optimism of those early days. Even in areas where success has indeed been achieved, human abilities have proved difficult to simulate. Consider chess again. Human players, whose thinking has been much studied by psychologists, appear to work with highly general, 'strategic' rules, such as one sometimes attributed to the former Cuban world champion Capablanca: 'Before launching an attack, close the pawn formation' (this means positioning your pawns so snugly to those of your opponent that neither player's pawns can capture the others, thus bringing a degree of stability to the game). Maybe one day computers will use strategic rules of this kind, but at present they don't. For them, chess is all tactics and no strategy. Essentially, they make most of their decisions by means of brute force search: going through all possible moves, all possible responses to these moves, and so on. Sophisticated search strategies invented for chess—which are able to selectively disregard certain moves—are a very substantial contribution to computing, with applications in many areas. Clearly, computer chess has achieved a lot. Still, some researchers are disappointed, because *people* could never play chess using search alone: even grandmasters cannot think far enough ahead to make it work. Conversely, computers would have great difficulty making sense of Capablanca's rule. The reason is that this rule is merely a rule of thumb. Suppose, for example, you can checkmate your opponent; closing your pawn formation first would be a blunder in this case. A rule of thumb is a *vague* rule: a rule that applies only most of the time.

The same pattern can be observed in much of AI: successes are achieved, but not overnight, and in unexpected ways. The use of ordinary language, on which we shall focus shortly, is a case in point. The trend is perhaps most evident in *machine translation*, which aims to let computers *translate* text from one language to another. Machine translation could, of course, be tremendously useful: if successful, it would amount to the demolition of the Tower of Babel! For some time, machine translation was seen as a potential 'killer application' for AI, but in 1966 a report commissioned by the American government concluded that machine translation was never going to work: the argument involved examples such as the following.

Suppose a computer needs to translate sentences from English to German. Suppose it reads the word 'open'.[2] How should this word be translated? If this word is posted on the door of a store then it means the shop is not closed: a German would say that it is *offen*. But if it appears on a banner in front of the store it is likely to mean something else: the shop has just been opened for the first time. In German this happens to be expressed differently, by the phase *Neu eröffnet* (i.e. 'newly opened').

The example hinges not on vagueness, but on that other pain in the linguistic neck: ambiguity. The problem is that the English word 'open' can mean two different things, while German does not have one expression covering both of these. Consequently, the only way to arrive at a good German translation of 'open' without guessing is by assessing how the English word was intended. The author of the report, a machine translation researcher named Yehoshua Bar-Hillel, argued that it would be so difficult for a computer to do justice to all the factors that can possibly be relevant for this assessment that it is simply not feasible.

Bar-Hillel's report was a blow for machine translation, effectively halting its progress for a long time. Forty years later, machine translation has finally recovered. Machine translation systems are starting to approach the point where they become commercially attractive. (Don't ask them to translate a poem though!) By and large, it is corpus-based methods (discussed in Chapter 6) that are making the difference. Current machine translation systems work by applying statistical methods to huge *parallel* language corpora, in which the same information is expressed in different languages. Broadly speaking, these methods figure out in which contexts a word such as 'open' tends to be translated as *offen* and in which contexts as *neu eröffnet'*. (The latter expression tends to co-occur frequently with such expressions as 'new', and 'location', whereas the other interpretation co-occurs more frequently with 'closed', for example.) Once again, an AI problem is starting to be cracked using techniques very different from the ones used by people.

After a lull in the 1980s, AI is once again thriving, though in a more modest spirit than before. Perhaps most importantly, researchers have come to realize that many things which *seem* easy are not. Playing chess may be challenging, but building a computer that can make conversation, recognize a person in a crowd, or play football, is proving to be even more challenging. It is, of course, rather humbling to realize that we share these 'problematic' abilities with little children. Perhaps we need to think differently about what it means to be human. If it is so difficult to let robots perform these 'non-intellectual' tasks, then perhaps we need to take them more seriously. It could consequently be argued that they should feature more strongly in our assessment of *human* intelligence (see Chapter 3).

Artificial intelligence?

Are computers intelligent? Arguably this question was thrown up in earnest for the first time by one of the founding fathers of computing, the Englishman Alan Turing. His work was so important and wide-ranging that, from where I stand, it is difficult to think of a more influential figure in the entire previous century.[3] Our understanding of the power and limitations of computers, for example, owes a tremendous amount to his work. His practical achievements are no less impressive: some historians believe that the Second World War would have ended differently without his contributions to code-breaking. In what follows, we focus on another, more speculative part of his work, involving what has come to be known as the Turing Test.

Thinking about intelligence, Turing was fascinated by a conversation game. The game is best played using a keyboard, so the participants communicate without hearing or seeing each other. In one version of the game, the role of a player we shall call the *deceiver* is to fool the other about his gender; the role of the player, whom we shall call the *detective*, is not to be fooled. The detective might, for example, start the game by asking what the deceiver is wearing. If the deceiver is a man, intent on making the detective believe he is a woman, he might lie and boast about his dress. (You haven't seen anything like it, an absolute bargain.) Being asked about the brand and cost of the dress, he should show decent knowledge of such things, or else the detective will see through his dastardly scheme.

For the deceiver to succeed, he has to think like a woman. Turing realized that the game could be turned on its head if the role of the deceiver is played by a computer rather than a person. The task for the detective in this case is no longer to determine the deceiver's gender—it doesn't even have any—but to find out whether the deceiver is a computer or a real person. The computer's task is to fool the detective into believing the latter. Analogous to the original game, the computer can win only by thinking like a human, to such an extent that, by observing the deceiver's contributions to the conversation, the detective cannot tell that the deceiver is a computer. Now suppose one carried out a test in which a computer was able to fool a majority of human subjects into believing it to be human. This would mean that people really cannot tell whether they are dealing with a person or a computer: we're just guessing. If this situation ever comes to pass, then surely the computer must count as human as far as its intellectual capacities are concerned, so Turing argued. If this ever happens the computer will be said to have *passed the Turing Test*, in which case one of the main aims of AI will have been achieved.

Turing's ideas about the Turing Test have sometimes been questioned, but the concept behind it is very much alive. In 1990, for example, the American philanthropist Hugh Loebner promised to award $100,000 for the first computer program to pass the Turing Test, and smaller yearly prizes are awarded to the program that comes closest to passing it. Additionally, a lot of other AI research focuses on small parts of Turing's challenge and can be seen as ultimately motivated by a desire to create machines that are intelligent in Turing's sense.

In the next chapter, we shall look in greater detail at AI researchers' attempts to get computers to pass the Turing Test. But first, let us review a few other areas where AI has had to deal with gradable phenomena. To start with, let us take a look at models of non-specialists thinking about physical processes.

Qualitative reasoning

The aim in *qualitative reasoning* is to construct computer models that focus on the essence of a process instead of its details. A model of this kind might say, for example, that the colour of your coffee will get lighter as you add more milk, without quantifying the effect. Let me explain how AI systems in this area work by looking at another example, involving a bathtub which is being filled with water.[4]

What happens if you open the tap and let water flow into the tub? Most of us would answer this question along the following lines: if the drain is closed then the level of the water will rise until it reaches the top. At this point the water will overflow and flood the bathroom. If the drain is open then the amount of water escaping via the drain will increase as the water level increases, and the tub will overflow only under certain conditions, relating to the amount of water flowing in, the size of the drain, and the shape of the tub. Physicists, of course, would describe the problem in much more quantitative terms. Most of us get by very well without so much detail, however, using a model that is merely *qualitative*.

Computer models of 'proper' physics are not difficult to construct, as long as the processes involved are not too complex. But if our aim is to model the knowledge that ordinary people bring to everyday tasks then a less detailed, qualitative model would be interesting to have. Such a model could be practically useful too, for example for educational purposes. A qualitative model of the human heart, for example, could help to explain the function of the heart to a medical student, elucidating the basic principles that are at work while suppressing unimportant details. Or suppose you were an inventor, playing around with innovative new heating systems: at the early stages of your work, you might benefit from a qualitative model that shows you which ideas might work and which ones won't. Details can obscure the things that matter.

What might a qualitative model of a bathtub look like? The basic behaviour of one famous program goes as follows, focusing on the situation in which the drain is open. The laws governing the system are simplified to the following regularities, where *Amount* is the amount of water in the bathtub, *Level* the level of the water in the tub, *Inflow* the amount of water flowing from the tap, and *Outflow* the amount of water escaping through the drain. *Netflow* is the net gain resulting from Inflow and Outflow.

1. Level goes up if and only if Amount goes up.
2. Outflow goes up if and only if Level goes up.
3. Outflow + Netflow = Inflow.
4. Netflow gets added to Amount.

The second of these regularities, for example, says that the amount of water escaping through the drain increases as the water level goes up, but *without* saying how one increase depends on the other. The program starts by activating a set of initial conditions which hold that, before the tap is opened, Amount = Level = 0. In this situation, the third law allows the system to infer from 0 + Netflow = Inflow that, at this moment, Netflow = Inflow. Since the tap is running, the fourth law implies that Amount must be going up. Now the first two laws kick in again to tell us that Level and Outflow go up as well. Since Inflow is constant, the third law implies that Netflow must be going down. The most interesting part of the model is reached when another law kicks in which says that several things can happen to the Level: either it may never reach the top of the bathtub, staying steady at some lower level because an equilibrium is reached, or it may reach the top. In the latter case, the water might either be overflowing, in which case we enter a different system of regularities, or just happen to remain steady at the top—an unlikely equilibrium, but not impossible.

Qualitative reasoning systems capture regularities while suppressing details. In ordinary language, the same function is often performed by degree adjectives:

> If a river flows fast mud and sand is carried out to sea. If a river flows slowly the mud sinks to the bottom—and makes the river shallow. The Clyde needed to be deep enough for ships, so walls were built to make the river narrower and faster. (Exhibit explanation in Clydebuilt, a museum devoted to Glasgow's industrial heritage)

When used in this way, there is nothing vague in words such as 'fast': the speed of a river is positively correlated with its depth, although nothing is said about the strength of this correlation. Similarly, the bathtub laws say that a higher value for Level implies a higher value for Outflow, and conversely. Nothing would be easier to express in classical logic. It is only if qualitative reasoning is combined with fuzzy logic that vagueness starts playing a role.

An area of AI where vagueness really does take centre-stage involves the construction of computer systems that give people *advice* about what to do in difficult situations: these systems offer decision support. To explain the role of gradability and fuzziness, we shall have to go into a fair amount of detail.

Applying fuzzy logic: An artificial doctor

Decision support systems are important in fault-critical domains such as medicine. Systems of this kind work with rules that give advice in various situations. These rules are often the result of extensive interviews with experts, but it is their combination that the computer program is responsible for. Doctors might, for example, inform the designer of the

229

system of the following rules that are relevant for a particular cardiac operation:[5]

Rule 1: *If* blood pressure is low *or* body temperature is low *then* risk is low.

Rule 2: *If* blood pressure is moderate *and* body temperature is high *then* risk is moderate.

Rule 3: *If* blood pressure is high *then* risk is high.

Rules of this kind do not state ordinary conditionals as they are used in classical logic. (If they did, then contradiction would loom if blood pressure is high but body temperature low.) Also, Rule 3 does not simply divide patients into the ones that do and the ones that do not have high blood pressure; rather, it suggests that, as blood pressure increases, so does the risk involved in the operation, other things being equal. A natural way to understand these rules arises if we take a fuzzy logic perspective. Let us see how this can be done.

Suppose, for simplicity, that Rules 1–3 are the only rules, so each of the three risk levels, low, moderate, and high, is affected by only one rule. Let us see how rules of this kind can produce conclusions about a patient. In doing so, we shall make use of one of the oldest proposals in this area, by Ebrahim Mamdani and his colleagues in London in the 1970s. Once the rules are in place and basic measurements have been taken, Mamdani-style inference proceeds in four steps, called fuzzification, rule evaluation, output aggregation, and defuzzification. Suppose the patient's blood pressure is 150/90 and her body temperature is 37.7 °C. (See Fig. 19, where the two figures that make up a blood pressure reading are averaged. Similar graphs are assumed to be available for body temperature and risk level.) Here is how her level of risk is computed.

1. *Fuzzification.* The first step is to 'interpret' the measurements taken from the patient in terms of the concepts featuring in the rules. The rules are cast in terms of blood pressure and risk being low/moderate/high,

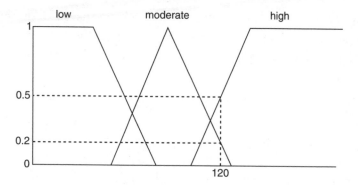

FIG. 19 Blood pressure in a complex membership function (average between systolic and diastolic values)

and temperature being low/high. The price to be paid for this subtlety is that degrees of truth must now be assigned to *each* of these three values: we have to say to what extent the patient's blood pressure counts as *low*, and to what extent as *moderate*, and to what extent as *high*. Let us suppose that, given her age and other factors, the patient's blood pressure counts as v(Pressure is high) = 0.5, v(Pressure is moderate) = 0.2, and v(Pressure is low) = 0. (See Fig. 19. Note that fuzzy logic does not require these values to add up to 1.) Suppose her body temperature is such that v(Temperature is low) = 0.1 and v(Temperature is high) = 0.7. In other words, our patient's blood pressure and body temperature are both on the high side.

2. *Rule evaluation.* The results of the previous step determine to what extent each of the three rules 'fires'. The most straightforward is Rule 3, which has only one precondition, that *pressure is high*. Since fuzzification tells us that v(Pressure is high) = 0.5, this rule fires to degree 0.5. As a result, the conclusion offered by this rule, that *risk is high*, also has a value of 0.5. The same technique is applied to Rules 1 and 2, but a few wrinkles need to be ironed out, because these rules involve *compound* preconditions, involving 'and' or 'or'. Rule 1, for example, has the precondition *blood pressure is low* or *body temperature is low*, whose first part has the value 0 and whose second part has the value 0.1. The value

231

of a disjunction equals that of the 'truest' of its parts, as we have seen, which is 0.1, meaning that this rule is only marginally relevant. Rule 2 is only slightly better off, because it fires to a degree of 0.2, the *minimum* of 0.7 and 0.2.

3. *Output aggregation.* We now know that the risk from the operation is *high* to degree 0.5, *moderate* to degree 0.2, and *low* to degree 0.1. This information is now combined into one complete membership function, which says for each risk level between 0 and 1 to what degree this risk level applies to our patient. As always, this information can be displayed in a membership graph.

4. *Defuzzification.* So how much is our patient at risk? Mamdani assumed that the take-home message for doctors should not take the form of a complex membership graph (as delivered by the previous step) or vague words such as 'high' or 'low', but a single *number*: a fuzzy truth value between 0 and 1. One way to find a suitable number is by drawing a vertical line through the graph of the membership function right where it cuts the area under the graph in two equal halves. The location of this line will tell us that, in the situation of our example, the risk is slightly above 0.5.

These techniques—which we have sketched loosely here, glossing over many mathematical details in the last two steps—represent only one way of doing things. At every step, sensible alternatives exist. This is perhaps easiest to appreciate in connection with *fuzzification*, but it is also true for *rule evaluation*. It is, for example, not obvious that we have done justice to Rule 2 by taking the minimum of the values 0.2 (for the degree to which pressure was moderate) and 0.7 (for the degree to which temperature was high) since, in the end, this failed to take temperature into account entirely. Complications with *defuzzification* arise if several rules address one and the same factor, in which case a way should be found to deal with accrual of evidence. (If three people tell you that the butler is the murderer, this should carry more weight than if only one person does.) In all these cases, alternatives exist. This is no coincidence: once truth values proliferate, then so do the possibilities for handling them. This is an embarrassment of riches that fuzzy logicians have to live with.

The future of AI

In middle age, after its first half century, artificial intelligence has become more like other science subjects. Exaggerated promises are seldom heard, except in the popular media, where robots go on conquering the world. A new realism has turned AI into a research area some of whose most application-oriented parts are closely linked with the computing industry, while its more theoretical parts are essentially an area of psychology. Debates about the limits of AI (Can computers be conscious? Can they have emotion? Could they ever become as intelligent as people? Do robots have rights?) flare up from time to time, but conclusions are not in sight. We leave these issues to better minds.

It would be ridiculous to suggest that *all* the obstacles facing AI stem from vagueness, but vagueness is implicated in a number of them. We have seen that this is true for computer chess, and it can certainly be argued in the case of AI work on communication, as we shall soon see in more detail. Other examples abound. Consider visual *object recognition*, for instance, which is the task of allowing computers to recognize particular types of objects, such as guns for example (in airport security), landmines (for mine clearance), or people. One of the most serious problems in this area is that the types of object in question are often difficult to define. Consider *people*, for example. The possible dimensions of a person are virtually impossible to state with any precision, and any attempt would tend to include false positives (plastic fashion dolls taken to be people) and false negatives (people who fail to be recognized as human because they are unusually thin). The challenges to automatic recognition of speech and handwriting, both of which are key problems in AI, are not dissimilar. It would be reasonable, in my view, to hope that a better understanding of vagueness might shed light on some of these other problems as well.

Given that the boundaries of artificial intelligence are only loosely defined, it is difficult to quantify the extent to which today's world is already applying AI techniques. Yet, some people have tried. A 2003 report by the Business Communication Co., for example, estimated the global market for AI at close to $12 billion. What is clear is that there are important applications in areas such as fraud detection (e.g. detection of credit card fraud), security (e.g. automatic analysis of email conversations), car navigation (e.g. route finding and description), and the medical domain (e.g. decision support and data-mining). Defence is probably the biggest single investor, reportedly with huge pay-offs: it has been estimated that the deployment of a single logistics support aid during the Desert Shield and Storm Campaigns against Iraq in 1990–1 paid back all US government investment in AI research over a thirty-year period. Whether these are the best possible uses of AI is a question I will not attempt to answer.

As for *communication* with computers, there is a long way to go before computers might ever be able to use language in a way that resembles human speaking or hearing, but progress is being made all the time. It is this area of AI on which the next chapter will focus.

Things to remember

◆ Artificial intelligence (AI) is an area of computer science in which computer programs are constructed that mimic human abilities. This can be done either for practical purposes (i.e. because the program performs a useful task) or because a computer program is an interesting model of a thinking agent.

◆ AI has become strikingly successful in recent years, particularly in areas such as computer chess, planning, robotics, and decision support.

In many cases, however, the most successful computer programs work quite differently from people. They may emulate *what* people do, but seldom *how* they do it.

♦ The drawbacks of fuzzy logic that were identified in our theoretical discussions of Chapter 9 turn out to plague practical applications of fuzzy logic as well. This became clear in our discussion of decision support systems, which could produce advice only once a programmer had decided how to define the logical connectives of the system.

♦ Qualitative reasoning systems perform automated reasoning with physical quantities by focusing on broad tendencies (e.g. a variable whose value increases, decreases, or stays the same) while disregarding quantitative detail. Because of their ability to suppress details, qualitative reasoning systems can provide interesting and useful models of people's understanding of physical and other regularities.

♦ Vagueness and a number of problems structurally very similar to vagueness are implicated in many of the greatest obstacles facing AI at the moment.

♦ Artificial intelligence has experienced a peculiar reversal: tasks that were originally thought to be hard (such as chess, for example) have turned out to be much easier to tackle than some tasks that are sometimes thought to be easy (such as interacting socially with one's environment). It can be argued that this experience should have implications for the testing of human intelligence, where more emphasis might be placed on *social* abilities than is commonly done.

♦ Communicating through human languages (such as English or Chinese, for example) is one of the AI tasks that are harder than anticipated. One aspect of human communication will be discussed in the next chapter.

11

When to be Vague
COMPUTERS AS AUTHORS

If I seem unduly clear to you, you must have misunderstood what I said.

ALAN GREENSPAN, President of the Federal Reserve Bank[1]

In earlier parts of the book, we have focused on the meaning and interpretation of vague language. In the present chapter, we shall ask when it is a good idea to be vague. As before, we shall often take a 'constructive' viewpoint, asking what it would take for a machine to use vagueness judiciously in its communication with people. It will therefore be instructive to see how computers use vagueness. But first we need to discuss what talking (or writing) computers can be good for in the first place and how they work.

Computer simulations of human communication are being used in a growing number of practical applications. Many of these arise from the fact that the world around us is full of information encapsulated in numbers and other symbols. Often this information is itself produced by computers, which measure and monitor and calculate things for us all the time. As this information grows, ever smarter solutions are needed to make it accessible and understandable. Examples abound, from sport and medicine to astronomy. We shall zoom in on one of the oldest applications of this kind, namely computational weather forecasting.

The expression of *numerical* information is paramount in this application: think of the measurement of temperature, wind speed, air humidity, and so on. Other applications may depart from different kinds of information, but the issues for AI are essentially alike.

Computer-based weather forecasting has been around since the middle of the 1950s. Forecasting programs produce huge tables of numbers about temperature, rainfall, humidity, wind, and so on. These numbers are only predictions, but they get better all the time, and they are considered reliable enough to be broadcast to the general public, as well as to seamen, air traffic controllers, and so on, who use them as a basis for decisions about serious matters. Unfortunately, the tables churned out by the computer are difficult to understand, because there is just too much data—and uninterpreted data at that. Luckily, the weather people understand that these figures need to be filtered and interpreted, leading to the kinds of forecast that you and I get to see. The task is essentially the same regardless of the area whose wheather is being forecast, and regardless of the language in which this is done. Here, for example, is yesterday's forecast for one part of Great Britain, in one of its national newspapers:

> SW England: There will be some hazy sunny spells around at first, with the chance of some scattered showers. Becoming cloudier through the afternoon with outbreaks of rain expected in the south by evening. Max temp 15–18C (59–64F). Tonight, rain. Min temp 9–12C (48–54F).

The BBC's flagship 10 p.m. television news programme concerning the same period was more elaborate, with plenty of colourful language. Here is an extract, recorded as spoken by a weatherman waving at a map; the forecast starts with some generalities about Britain as a whole, followed by a chronological narrative.

> A really quite decent day for many parts of the British Isles. (...) There are lumps of cloud, the odd rumble of thunder too. (...) Essentially quite a mild and quite a dry night too, for all but those in the glens of

Scotland. (…) It's a dry enough start, but it will be really quite breezy. (…) Enough cloud in the heart of Wales for the odd light shower. It's not wall-to-wall sunshine by any means at all, but at least it's dry and it's fine, the on-shore breeze just making it a wee bit chilly, as it has done for the past couple of days. There is some sunshine to encourage you out and about. (…) There are one or two showers to be had as well across the heart of England and Wales. Best of the sunshine through central and western Scotland and Northern Ireland.

Media communicate the weather forecast in their own style. But they invariably 'translate' numerical information into words. There tend to be additional numbers visible on a map, mainly about the air temperature. Information about wind speeds, however, is conveyed through such expressions as 'really quite breezy'. This makes good sense, because those of us who are not sailors may not be familiar with wind numerical metrics for speed. The words do not just *omit* information: by describing the wind speed as they did, the weather people tell us that there is more wind than usual, but there is no cause for concern.

It has been said that the weather is Britain's chief religion these days, which is why it is such a mainstay of everyday conversation there. In such a delicate area, we must tread carefully. Yet, what would be more natural than using computers to take away some of the work from the forecasters? Given that computers are used to produce numerical forecasts, it's natural to use computers for the next step—from numbers to text—as well. This is particularly useful when weather reports are needed for a specific small area, such as oil rigs out at sea, which lack the resources to produce dedicated weather reports by hand. For reasons like this, weather forecasting is one of the oldest applications of *natural language generation* (NLG), which is the area of AI that focuses on the expression of numbers, or other symbols, in ordinary words.[2]

A good example of an early system of this kind is the aptly-named FoG system,[3] which automatically generated written summaries, with little else than numbers as input. FoG produced brief, businesslike weather predictions, focusing on weather features such as wind speed, which are particularly relevant for marine operations. FoG generated gale warnings, for example, from input such as 'windspeed(morning) = 37 knots'. The use of expressions such as 'gale' or 'strong wind' was governed by a simple set of fixed thresholds. Gale force winds, for example, were defined as being greater than 32 knots and less than 48 knots. Many AI systems, in a wide variety of applications, work in essentially this way, categorizing numerical data into a few broad classes, which are separated from each other with almost absolute crispness.

This approach, however, is prone to errors. If a programmer, on a sleepy afternoon, enters the wrong threshold, deciding that 20 knots counts as gale force, there is nothing to stop him. The problem becomes worse if the system needs to be sensitive to circumstances. Suppose, for example, you want a particular threshold to depend on the season. In Scotland, for example, 13 °C is often described as *mild* on a winter day, as *seasonal* in April, and as *cold* or *disappointing* (as if they knew what disappoints us!) in August. The picture is complicated even further because we have different expectations about temperatures at different times of day. To enter different thresholds manually for all these different situations would be a nightmare. The recommended alternative is to take note of linguists' work on vagueness, and to devise a formula that *computes* the threshold in each different situation, depending on the time of day and the time of year. Newer systems are moving in this direction. Their output is now good enough that, in certain niche applications such as local forecasts for individual oil platforms at sea, it is starting to be economically viable to let an artificial weatherman replace the real one. In applications of this kind, computer-generated texts can sometimes be more effective than human-generated ones, because of their predicability. We saw in Chapter 7 (in the discussion of vagueness as ignorance) that it can be strikingly unclear what a human forecaster means when she says that, for example, rain will start 'by evening'. Computer-generated texts reduce this problem considerably because, at least, it is easy for them to be perfectly consistent in their usage of words.

More generally, the presentation of quantitative information is a lively area of research. Researchers from different disciplines collaborate to find out how complex facts can be expressed transparently in ordinary language, and how such transparent explanations can be automatically generated by computer, based on nothing more than numbers. Deciding how to explain numbers can require considerable psychological insight, however.[4]

Medical applications figure prominently in this area. Suppose you take tablets against hayfever and you hear that the risk of certain unwanted side effects—let's say a heart problem—is 0.0001. Is this a reason to avoid the tablets and keep sneezing? Bertrand Russell once wrote that people have a tenuous grasp of anything that is unusually large or unusually small: we can only deal with things whose sizes lie within some relatively small human comfort zone. For most of us, the risks of medical side effects tend to lie well outside this comfort zone. To compound matters, there is the complication posed by *relative* probabilities. Many of us have difficulty realizing that if, say, diabetes makes you 70 per cent more likely to develop some very rare disease, this does not mean that a person with diabetes has a high likelihood of contracting the disease. The research suggests that even doctors struggle interpreting relative probabilities correctly.

It is widely recognized that words can help to make numbers transparent. In recognition of this fact, the European Union has published official guidelines for warnings of side effects. These guidelines recommend using the expression 'very common' for probabilities in excess of 10 per cent, for instance. Research suggests, however, that the general public understands this expression as indicating a probability of as much as 65 per cent on average. More generally, people vastly overestimate the probabilities associated with each of these expressions. To top it all, it has been shown that different people interpret the words in question in wildly varying ways, so there is little hope of solving the problem by simply replacing one word with another: a description that would give one person the right idea would thoroughly mislead another. Words are not always preferable to numbers.

Perhaps the way forward is to look for suitable comparisons. Instead of letting the computer say 'you have a chance of 0.0001 of getting disease X', it might be better to say: 'In a town of 1,000,000 people, I would expect 100 to get X.' And, instead of saying, rather abstractly, that your risk of contracting X goes up by 70 per cent as a result of having diabetes, it is better to say that if no one in this town of 1,000,000 people had diabetes, you would expect (for example) 100 people to get the disease, but if they all had diabetes your expectation would increase to 170. Graphical depictions can help as well, because the size of a graphic can be used to show the size of a number: you could use a square of 100 × 100 mm to depict a population of 1,000,000, and a smaller square of 1 × 1 mm to depict a group of 100, for example. This way, the size gap leaps out at you.

Two lessons should be learned: first, finding a good English (Dutch, Chinese, etc.) explanation for complex information may require considerable invention. Secondly, ordinary language may not be able to do the work on its own: other media may offer an essential addition. These lessons will be kept in mind when we examine one area of natural language generation (NLG) in more detail.

Example: Generating vague descriptions

Quantities are expressed in many contexts and for many purposes. To make things more concrete, let us look at one specific NLG task which is relatively well understood. The task I have in mind is *reference*.[5]

Reference occurs when we draw someone's attention to an object. We can do this through a gesture, for example, or through words. Proper names, such as 'Mount Everest' or 'Albert Einstein', are often used for this purpose, but there are many situations where proper names are not available. Not many sparrows have proper names, as far as I know, and the same is true for knives and forks. Even my own left arm lacks a proper name. Bacteria are no better off, nor are all except a small minority of stars. Come to think of it, proper names are a rare privilege.

In the absence of proper names, how can a thing be identified? Consider a fork that you have just used. I might refer to it as 'the fork', 'the silver fork', 'the large fork', 'that thing over there', and so on. Each of these references express a combination of properties (being large, being made of silver, and so on). If I choose them well you know which object I'm talking about; if my choice is clumsy I may be giving you a hard time: you may even misunderstand which fork I'm talking about. When computers perform a reference task, their choice is based on a search of all the properties in a database, coupled with some simple assumptions about the way in which hearers interpret language. They usually operate under a constraint that is typical for much work in artificial intelligence, as we saw in the previous chapter. They aim to refer in a human-like way: in such a way that they might pass the Turing Test, in other words.

Reference is one of the most widely studied areas of communication, with important contributions from linguists, psychologists, and computer scientists. Yet until recently, the NLG programs that perform reference never took vagueness into account: the properties they were looking for were always crisp. Moreover, they were never context-dependent. To see why this is a problem, let us look at an example.

Suppose an NLG program operates in a kennel where some fancy AI software has been installed. Every dog is classified as a particular type of dog (alsatian, poodle, etc.) with a particular fur colour (brown, black, you name it) and size (e.g. large or small). Now suppose a new kennel member arrives: a chihuahua. Chihuahuas, as a general rule, are smaller than other dogs, but this particular one is larger than usual: larger than quite a few poodles in fact. Still, it is much smaller than most other dogs, so it is listed as *small*. Dogs never stay in the kennel for very long, so the fact that they grow can be neglected.

Ann arrives at the kennel, keen to buy a dog. She does not have strongs views on the size of a dog, as long as it's not ridiculously small.

242

Chihuahuas would be great, just not the smaller ones. Ann sits down behind a PC on which a user-friendly, artificially intelligent interface has just been installed. After some browsing, the system comes up with suggestions, in proper English no less. The first suggestion says 'small chihuahua', in an attempt to refer to our friend, the new kennel member. That's a pity, Ann thinks. A good-size chihuahua would have been terrific, but not a small one please. (Bob would feel ridiculous walking it in the park.) As a result of this misunderstanding she might be missing out on the dog of her life. Something has gone wrong here, for the kennel *did* have a dog exactly to her liking. But because the poor animal was misdescribed as a 'small chihuahua', Ann mistakenly thought it was small *for a chihuahua*. The animal was small for a dog, but not for a chihuahua!

This sad story demonstrates why some properties should not be treated as if they held independent of the context in which they are used: words such as 'small' are context-dependent, as we have seen in Chapter 6. The problem is that the clerk who enters dogs in the database has no way of knowing in what context its properties will show up in the user interface: that, after all, is the result of some smart language-generating program whose behaviour might change with every new kennel member, and with every new update of the system.

Instead of the somewhat dodgy system operated by the kennel, a really fancy interface would always be clear and understandable. If the program uses its language in anything like the way people do—a necessity if the program is ever to pass the Turing Test—then it will be able to refer to one and the same dog in many ways: 'The large chihuahua', 'the small dog without a tail', and so on, always selecting the description that fits the situation best. Systems of this kind are one of the holy grails of NLG. Ideally, programs of this kind should be completely general of course: they should work not only in connection with dogs and their sizes, but equally in connection with digital cameras or second-hand books and their prices, for instance.

How can a program of this kind operate? First of all, the facts themselves should be kept neutral between all the relevant contexts. Instead of saying that a given dog is *old*, for example, the clerk should list its age, in years perhaps. Similarly, the size of each dog will be recorded, perhaps in centimetres. All this is added to the usual properties, such as type, colour, and so on. When the clerk is done, the database might look like this, with four chihuahuas, a poodle, and a bunch of alsatians thrown in for good measure:

$$\text{Type}(c_1) = \ldots = \text{Type}(c_4) = \text{chihuahua}$$
$$\text{Type}(a_1) = \ldots = \text{Type}(a_{10}) = \text{alsatian}$$
$$\text{Type}(p_5) = \text{poodle}$$
$$\text{Size}(c_1) = 12\,\text{cm}$$
$$\text{Size}(c_2) = 20\,\text{cm}$$
$$\text{Size}(c_3) = 30\,\text{cm}$$
$$\text{Size}(c_4) = \text{Size}(p_5) = 38\,\text{cm}$$
$$\text{Size}(a_1) = \ldots = \text{Size}(a_{10}) = 80\,\text{cm}$$

Suppose c_4 is the animal that needs to be singled out. The properties in the database are examined by the program in the order in which they are listed, until the referent has been singled out clearly. Applying this recipe to our small kennel, the following will happen. The first property attempted is 'chihuahua'. Since not all the dogs in the kennel are chihuahuas, this property can be a useful component of the description of the dog and will therefore be kept, removing the poodle and all the alsatians from consideration because they're the wrong kind of dogs. Now the size of the dogs is taken into account. The only size applicable is the property $\text{Size}(x) = 38\,\text{cm}$ and, in combination with the property 'chihuahua', this one happens to single out c_4. The resulting set of properties is

{Type = chihuahua, size = 38 cm}

This might be considered the end of the matter, since the target object has been singled out by means of a suitable set of properties, and a good referring expression could be generated from this set without too much difficulty: 'the 38 cm chihuahua' for example, or 'the chihuahua whose size is 38 cm'. But we want to accommodate Ann, who may not have a clear idea of these size measurements. Also, she might not be fluent in the language of centimetres, preferring feet and inches instead. For her benefit, we want to generate a referring expression containing a vague adjective: Something like 'the large chihuahua' would do nicely, for example.

Given that 38 cm happens to be the greatest size of all the available chihuahuas, $Size(x) = 38$ cm can be replaced, in the set of properties, by the property of 'being the largest member of' (notation: $Size(x) = max$). This property is applicable only because of the properties introduced earlier into the description.

{type = chihuahua, $Size(x) = max$}

This list of properties can now be put into words as, for example, 'the largest chihuahua', 'the larger chihuahua', or 'the large chihuahua'. The difference is an issue on which a considerable amount of empirical research has been conducted, but which we shall leave aside here because it would distract us from the core of the generation procedure.[6]

It is worth reflecting a little on what has just been done. A vague description has been used to describe an animal. Or has it? On the one hand, 'large' is as vague as a word can be, but on the other hand, it has been made part of a description that can be interpreted with absolute precision by anyone who knows the kennel. Suppose someone—not Ann, alas!—walks up to the clerk and says 'Can I have the large chihuahua please?' One look at the database (or at the dogs themselves) would suffice to pick out the animal. The vagueness of the word disappears, as it were, because of the way in which it is applied. In an utterance such as this, 'large' is *locally* vague, but *globally* crisp, because when the utterance situation and the words around it are taken into account, the vagueness disappears.[7] We might also say that the word is only *apparently* vague.

Admittedly, this story makes some important simplifications. Suppose the chihuahuas in the domain are much closer together in terms of size, perhaps 12 cm, 12.5 cm, 13 cm, and 13.5 cm. In this situation it would hardly be natural to refer to the largest one as the large chihuahua; if they are even closer together—as in a typical sorites argument—then the expression would not even be understood, because Ann would not have been able to tell which one was largest. Other complications arise if we want to refer to *sets* as well as as individuals. Perhaps more interestingly, one could wonder what happens if a dog is identifiable through either its height or its girth or any other two gradable dimensions. Experiments have shown that human speakers prefer the description that utilizes the most striking difference in such cases: if the tallest chihuahua is also the fattest, but only by a small margin, then it's the 'tall' chihuahua, for example, not the 'fat' chihuahua.

Perhaps the hardest problems in this area arise from a source of vagueness that is always with us. This source of vagueness is known as *salience*. In many situations, salience helps us communicate. Consider the example of the fork again. How can we ever hope to identify an object by saying something as simple as 'the fork', or even 'the silver fork'? There must be billions of forks in the world, and a good many of them will be silver ones as well. The reason why we often get away with a short description such as 'the fork' is that sensible listeners will assume you mean 'the most salient fork'. Similarly, 'the silver fork' means 'the most salient of all silver forks'. Salience is essentially a psychological concept, to do with the degree of prominence of an object in the mind of a listener. Objects near you will tend to be more salient than objects further away. Objects that have recently been mentioned are more salient than others. Brightly coloured objects tend to be more salient than duller ones, and so on. One problem with salience derives directly from its being composed of different factors (proximity of the referent, recency of its being mentioned, etc.) For how should the different components of salience be weighed?

The situation can be illustrated with a problem that plagued my colleagues and me when we worked in a research lab in Brighton, near a small railway station. If someone said 'You need to be at the railway station at 7 p.m.', they could either mean this nearest station ... or the much busier one further away, in the centre of town. Misunderstandings were always around the corner. Essentially, our problem was that the salience of a railway station is determined by different dimensions and it was unclear to us how each of these should be weighed. Each of the two railway stations was more salient than the other in one respect, and less salient in another, and this kept creating confusion.

To sum up, salience is multidimensional, and this can cause problems. But suppose, for a moment, that all these dimensions could somehow be collapsed into one objective metric that told us how salient an object is at a given moment. There would still be the question how salience should be balanced against *other* gradable concepts. To see what is at stake, let us return to our earlier example, which had Ann visiting a dog kennel. Suppose our desirable friend, the chihuahua c_4, is much further away than all the other dogs, while c_3—the next largest one—is particularly close by. Under these circumstances one might say 'the large chihuahua' to refer to the latter, in recognition of its being near and large. But what if c_4 is *a little* nearer, and therefore more salient? Or what if it has also just been mentioned (which would have the same effect)? In these cases it is not obvious which of the two dogs is more salient. For which of the two carries more weight: size or salience?

Much more could be said, but let us stop here to return to the big picture: gradable concepts can help NLG programs to identify objects. It might be thought that the approach that was taken by the computer program sketched above is limited to gradable adjectives, but this is not the case. Nouns such as 'girl' (where age is gradable) and 'friend' (where the degree of friendship can vary) can be treated in the same way. The story does not end with such obviously gradable cases. As we saw when discussing *prototypes* in Chapter 6, psychologists have long known that even presumably crisp nouns such as 'shoe' or 'window' or 'academic' can apply in different degrees, and this insight has a direct bearing on the things we are talking about at the moment.

Imagine a gathering of four people: a famous professor (a), a junior lecturer (b), a Ph.D. student working on his dissertation (c), and a policeman uncontaminated by academic life (d). It is unclear how many academics are gathering, since each of the first three individuals (but not the policeman) could be regarded as an academic, while each of the latter three (but not the professor) could be regarded as a non-academic. Because each of the group is an academic to a different degree, each of the following expressions can be used with approximately equal justification, and each of the three can be generated without difficulty using the method outlined above:

1. the academic (to refer to a)
2. both academics (to refer to a and b)
3. the three academics (to refer to a, b, and c)

Mercifully, words do not have to do all the work: people use body language to add redundancy to otherwise risky references. In recognition of this fact, work on reference is starting to move beyond words, taking body language into account: a pointing gesture, or even just a nod of the head, can help to tell your audience what you are referring to, sometimes making words superfluous. Once again, vagueness abounds. Suppose I single out my mother in a crowd, by pointing at her and saying to you, 'the lady with red hair'. Unless I touch her, my gesture is likely to be vague, in much the same way as the expression 'red hair': you follow the direction of my index finger, but from that alone it is impossible to tell who I am trying to identify. My gesture identifies a circle with fuzzy contours which includes some people and excludes some people, but leaves you uncertain about others. The pattern is familiar from vague words, and can be modelled in essentially the same way. This may be a healthy reminder that vagueness is not an accidental feature of the languages we happen to speak, but inherent in communication, regardless of the medium.

DIALOGUE INTERMEZZO
WHAT USE IS A THEORY?

Susan and Jennifer have just made it onto the plane from San Francisco to New York. Having caught their breath, they start talking. After exchanging pleasantries,

248

they start debating their respective roles in the meeting on an artificial intelligence project from which they are both returning. The plane is shaking as it approaches a low-pressure zone.

SUSAN: I have very little time for all that theoretical stuff.

JENNIFER: What do you mean? The other people on this plane would say all we do is theoretical.

SUSAN: Yes, but I don't. I get very worked up when someone wastes our time blabbering on about yet another model of vagueness. Another paper full of formulas—who needs it?

JENNIFER: It depends on the formulas I suppose.

SUSAN: You don't mean to defend the behaviour of the tall chap, this Dutchman, do you? We're meant to build a weather forecasting system, not some wretched Rube Goldberg machine that won't be of any practical use.

JENNIFER: And you know how to build such a system, I take it?

SUSAN: Broadly, yes. The only thing I don't have enough of is data. Suppose the wind speed is 40 miles per hour. I'd like to know how this is typically described by weather people: do they call it a strong breeze, very strong, gale-force, or what? Or take temperature, which depends on place and time: a warm day in Alaska is not the same as a warm day in Florida.

JENNIFER: Glad to hear you don't know everything.

SUSAN: Don't be facetious. This is not about knowing everything, but about knowing enough. I need to know enough to tell our programmers what thresholds to use when they categorize temperatures, wind speeds, barometric pressure, and all the rest of it. And that's not something these theoretical people are going to tell us. You know the saying: your systems' performance will improve with every theoretician you fire. (Here the pilot interrupts their conversation: Heavy weather is predicted, with strong gusts of wind. After a while the two scientists resume their conversation, apparently unperturbed.)

JENNIFER: Suppose you had all the data, what would you do?

SUSAN: I'd run some machine learning programmes on it to let the computer figure out how the words in the weather talks depend on the actual temperatures. If time allows, we'd have more data, taking the season into account and that sort of thing.

JENNIFER: Yes, but satisfying it isn't. It doesn't explain why forecasters express themselves the way they do.

SUSAN: Fair enough. I can see why you're interested in theory. But what beats me is all this stuff about indistinguishability. If your system can measure a difference then there is a difference. If not then, for all practical purposes, there is none. You simply use some well-chosen thresholds to decide what counts as warm, hot, and so on.

JENNIFER: That makes sense if all you want is to produce clear and informative weather reports. I would like our programs to talk like real people. I want them to pass the Turing Test...

SUSAN: (laughing mirthlessly) Pass the Turing Test? And you think you need this stuff about indistinguishability to do that?

JENNIFER: Here's an example. Suppose you're talking about the weather in California, first about the north, then the south. (The plane is entering an area of quieter weather. The shaking has diminished.) Now suppose, at that time and place, the threshold separating 'warm' and 'very warm' is 25 °C...

SUSAN: Fair enough.

JENNIFER: Now suppose the temperature measured in northern California is 24.99 °C, and the temperature in southern California is 25.01 °C. This will cause the program to classify the north as warm, and the south as very warm. Pretty unsatisfactory if you ask me.

SUSAN: But who will notice? You said your criterion is the Turing Test.

JENNIFER: Here the weather program goes: 'Ladies and gentlemen, it's warm in the north, and very warm in the south, with temperatures reaching

25 °C everywhere across the state. This weather talk was offered to you courtesy of Susan...'

SUSAN: *OK, maybe you have a point. But surely there are bigger issues for us to tackle, such as making sure our weather talks are enjoyable as well as intelligible. Mind you, we're talking as if human weather forecasters were infallible, but they're not: they've been shown to be inconsistent and unintelligible! Any decent computer-based forecasting system is going to outperform human forecasters in terms of the clarity of the texts that are produced. Believe me, our problems lie elsewhere!*

JENNIFER: *(sighs) You are a business person who happens to work in an area where AI has made important contributions. I am a scientist interested in human behaviour. Never the twain shall meet.*

Tolerance revisited

Artificially intelligent programs tend to deal with vagueness in rather primitive ways. One of their limitations is, almost invariably, to neglect the role of context. This was the case, for example, for many weather forecasting systems which were unable to associate such words as 'warm', 'pleasant', or 'disappointing' with different thresholds in different seasons or climates, or at different times of the day. We have also sketched how AI systems are starting to cope with some of these subtleties.

It is perhaps less obvious why AI systems should try to do justice to the intuition behind the tolerance principle. What would be wrong with a weather forecasting system that used the word 'warm' in different ways under different circumstances, but always crisply? Such a system might use different thresholds for day and night, different ones for each season, and so on, but these thresholds would always be crisp, that is, Boolean. Would its crisp handling of vagueness cripple the system, causing it to fail a Turing Test, for example? So as not to give the system an unfair disadvantage, we should give it a wide variety of different expressions to choose between, associating different sets of boundaries with 'a bit warm', 'on the warm side', 'mild', 'warm', 'very warm', and so on.

The answer as indicated by the dialogue intermezzo above, is that a 'classical' system of this kind would fall through whenever a small difference was interpreted as a step change. This could happen quite dramatically, for example, if the temperature oscillated around some significant boundary, such as the one between warm and very warm weather. In such a situation it would be misleading to say that the temperature alternated between warm and very warm. It would be much better to say that the temperature was stable, oscillating around, say, 25 °C.

It is time to step back from concrete applications, and to ask more systematically when it is a good idea to speak vaguely. The answers to this question will be as relevant to speaking computers as they are to speaking people.

A game theory perspective

Scientists interested in the meaning of language have long seen symbolic logic as their mathematical conscience. In Chapters 8 and 9, we have seen this programme in action, when we observed how logic can shed light on language. Symbolic logic is suitable for formalizing ideas, such as equivalence and logical consequence, which revolve around the notion of *truth*. But there is more to language than truth and falsity. First and foremost, language is an instrument for communication, and communication can serve all manner of goals. When I greet you, or offer you something for sale, what does this have to do with truth or falsity? It can be argued that, at the most general level, the key function of an utterance is its usefulness, also called its *utility*, understood in the widest possible sense. We saw an example of such an analysis in Chapter 7, in connection with Rohit Parikh's story about Ann's topology book, when we tried to understand how much the utterance 'It is blue' helped her husband to find it. Let me briefly return to the example of a weather forecasting system to show how considerations of this kind can help an NLG system.

Suppose a weather forecasting system produces English texts to inform the decision which roads will be treated with salt (i.e. gritted). The input to the program comes from a road ice simulation model, which produces a huge table (worth several megabytes of data) containing numerical predictions of the air temperature, road temperature, wind speed and direction, and so on, for parts of Scotland. There can be thousands of dangerous roads on a given night, and it is often not feasible to say *exactly* which roads are dangerous. The generator might approximate the data by offering one of the following summaries:

(1) Roads *above 500 metres* are icy.
(2) Roads *in the Highlands* are icy.

Let us look in a bit more detail at these two summaries. Even if both of them give reasonable indications as to which roads are actually expected to be icy, it could matter a lot which of them is generated, because each summary will lead to a different set of roads being treated with salt. In situations such as this, it is often useful to distinguish between false positives and false negatives: the former are roads that will be gritted unnecessarily, the latter are dangerous roads that will be left ungritted. We might find that the first summary has 100 false positives and only 2 false negatives, while the second summary—which describes a somewhat smaller area, we assume—may be more evenly balanced, with 10 false positives and 10 false negatives.

The decisions inherent in these summaries involve a trade-off between two kinds of cost: safety on the one hand, and money and environmental damage, through salt, on the other. In situations of this kind, *decision theory* would be a natural framework in which to compare the utilities of the two summaries, for example by adding up the two kinds of cost. If a false positive has a negative utility (i.e. a cost) of -0.1 and a false negative one of -2, for example, then the first summary wins the day, because although it has far more false positives than the second one (at 100 against 10), false positives count for so little that the relative paucity of false negatives in the first summary compared to the second (2 versus 10) tips the balance in favour of the first. Needless to say, the figures are crucial, and tricky to justify. What is worse, after all: damaging the environment by gritting 100 roads unnecessarily, or failing to protect ten drivers who are entering dangerous territory? The formula used above might be underpinned in various ways (e.g. by estimating how road gritting affects a driver's chance of having an accident), but the figures that were chosen here are essentially fictitious.

It is possible to focus on decisions of the speaker, but it takes two to tango. If we really want to understand communication, we need to look at the receiver (i.e. the hearer) as well as the sender (i.e. the speaker). If a sender chooses to communicate by email, for example, whereas the receiver is waiting for a phone call, then communication can fail. It is for reasons of this kind that *game theory* is often a more suitable instrument than decision theory, because game theory analyses how the actions and strategies of several individuals combine. Game theory was conceived in the 1940s (von Neumann and Morgenstern 1944) and has since come to be used extensively by economists, sociologists, and biologists. Linguists are queuing up to do the same.[8]

Let us see how game theorists define their own trade. In the words of Eric Rasmussen (2001: 11), 'Game theory is concerned with the actions of decision makers who are conscious that their actions affect each other.' A game, in this context, is little more than a closely regulated set of conventions within which *players* (i.e. the decision makers) can choose from a limited set of *actions*. These actions are the moves in the game. Combination of players' moves is associated with a *pay-off* for each player, which says how good the situation is for the player in question. There will often be things that a player does not know: she may not know, for example, which moves the other players will choose, and she may not know certain things about the world that may have an effect on the pay-off associated with an action. An important assumption typically made in game theory is known as *rationality*: every player seeks to maximize her own expected utility. Because one player's gain can sometimes mean another one's loss, game theorists look for stable combinations of strategies (know as equilibria) that give none of the players an incentive to switch to another strategy. The rationality assumption may not always be justified, but without it, it becomes harder to predict the outcome of a game.

In recent years, game theorists have paid much attention to the effects that communication and information have on people's decisions when not all players possess the same information. As a result, a new community of researchers has started to think about communication and language: it is no longer exceptional for economists to publish in linguistics journals, for example. Let us see what this new line of work can tell us about vagueness.

We start with communication in its simplest form, based on a type of game that philosophers such as David Lewis, and economists such as Crawford and Sobel, have started applying to language. Two generals, G_1 and G_2, are both intent on attacking an enemy. But while each general individually is weaker than the enemy, they can beat him if they attack at the same time. Communication ('I am going to attack now!') can help the generals to cooperate and win the battle. Nothing hinges on the belligerent nature of the example, because something similar happens when you try to meet a friend: neither of you may care where and when you meet, as long as you end up in the same place at the same time; communication can help you achieve this goal.

Games are often analysed using diagrams like the following, where each general can follow an attack or a no-attack strategy, leading to a total of four possible combinations of moves. The first number between the brackets shows the pay-off for G1 and the second number the pay-off for G2.

	G2 attacks	G2 does not attack
G1 attacks	(1, 1)	(−2, 0)
G1 does not attack	(0, −2)	(0, 0)

If this is the entire story then the generals can only gamble: with nothing to go by, a general might estimate the chance of the other one attacking at 50 per cent, which would make his expected pay-off upon attacking the average between 1 (if the other general attacks as well) and −2 (if he does not), which is worse than the safe pay-off of 0 that would result if he did not attack. But if the generals communicate then everything changes: they will no doubt attack, after notifying each other of their plans, since this gives each of them a better pay-off than before. Neither of the generals has an incentive to change his strategy unilaterally, since this would diminish his pay-off from 1 to 0. One can easily see this from the table: starting from the situation (1, 1) in the top left, two unilateral changes are possible. One involves a change by G1, from (1, 1) to (0, −2), which will set him back one point; the other involves a change by G2, from (1, 1) to (−2, 0), which will set G2 back one point. Because neither general has an incentive to spoil the party, this is an equilibrium.

This example shows the spirit of this type of game theory model. But how about the peculiarities of language that this book is focusing on? It seems that, in terms of utility, *ambiguity* and *vagueness* only pose a hazard. If one general promises vaguely that he will attack 'on Sunday' (leaving the week unspecified), or 'soon', then disaster strikes if the generals attack on different days. Reflecting on examples of this kind, it is difficult to see how ambiguity or vagueness can ever be beneficial. But then why is human communication so full of these phenomena? Are they just the result of sloppiness, like typos that are erased from a text as soon as they are recognized for what they are? Or can we think of situations where ambiguity and vagueness *increase* utility? A positive answer focusing on ambiguity was given in 1994, when two game theorists, Enriqueta Aragonès and Zvika Neeman, showed how ambiguity can benefit a politican who wants to be elected while also keeping his hands free in case he does get elected.

Suppose two unscrupulous politicians position themselves for an election. Not burdened with any convictions, they are free to choose between three different ideologies (left, right, centre), depending on which gives them the highest utility; additionally, they can choose between two *commitment* levels, c_{high} and c_{low}, both representable as numbers with $c_{high} > c_{low}$. For just now, one might think of these as a more and a less extreme version of their chosen ideology.

What combination of an ideology and a commitment level should each politician choose? This depends on the electorate, of course, as well as on the choices that are made by the other politician. Suppose there are three blocs of voters, V(left), V(right), and V(centre), corresponding with the three available ideologies. A leftist voter prefers a leftist politician, and preferably one with a high commitment level. Confronted with a choice between two right-wing politicians, our leftist voter will prefer one with a low commitment. A right-wing voter behaves as the mirror-image of the leftist voter, while the neutral voter is neutral between the two idiologies but, weary of ideology, she prefers low commitment over high commitment. Commitment, in other words, is relevant only for a choice between politicians of the same ideology.

If this is the whole story then politicians will choose an ideology and commitment level based on their estimates of the numbers of voters in each bloc, trying to maximize their expected pay-off, formulated solely in terms of the likelihood of winning the election. The task for game theory is to work out what *combination* of strategies might give these politicians the best combination of pay-offs, under certain assumptions (such as politicians' beliefs about the numbers of voters in each bloc).

But Aragonès and Neeman's model allows politicians to look beyond the election towards their anticipated time in government. What if an economic recession comes along, making it impossible to keep all one's election promises? Surely, a low commitment is easier to fulfil than a high commitment, so it is better to be elected on a low-commitment platform that does not tie his hands too much! To model this, Aragonès and Neeman formulate utility in a way that multiplies the probability of a politician winning the elections with a constant that is negatively correlated to his commitment level. So if a politician wins the election, a low commitment gives him a higher overall utility than a high commitment. Under these assumptions one can show that a low commitment level (i.e. c_{low}) can sometimes give a politician a slightly *lower* probability of winning the elections (because his core voters will be less inclined to vote for him), yet a *higher* overall utility, because his time in office will be easier if he gets elected.

It is sometimes thought that Aragonès and Neeman's model demonstrates how ambiguity can be used strategically, but that it fails to shed light on vagueness. It seems, however, that theirs is as good a model of vagueness as it is of ambiguity. Let me explain why.

Suppose the ideology in question—a leftist, or perhaps a populist one—is to take away money from the richest 10 per cent of the people and give it to the poorest 10 per cent. Commitment level, in this case, could be a way of making explicit what percentage of the income of the top 10 per cent to give away. One position might be to hold that this has to be, say, 50 per cent of their income, while another position might be to put this figure at 5 per cent. But if we identify high commitment with the 50 per cent position and low commitment with the 5 per cent position then neither of the two commitment levels would be ambiguous. To make one of them ambiguous, we need something else. For example, the higher of the percentages mentioned above can represent one strategy (indicated below as c_{high}), while the other one may be characterized by a more ambivalent commitment level ($c_{ambiguous}$).

The ambiguous politicians' game:

- Ideology: take money from the richest 10% of people and divide it equally between the poorest 10%.
- c_{high}: do this with 50% of the money of each of the richest people.
- $c_{ambiguous}$: do this with *either* 5% *or* 50% of the income of each of the richest people.

But this must be a simplification; there is nothing to exclude percentages between 5 per cent and 50 per cent, for example. It seems, therefore, perfectly possible to construct a version of Aragonès and Neeman's game that hinges on vagueness:

The vague politicians' game:

- Ideology and c_{high}: as in the ambiguous politician's game.
- c_{vague}: carry out your ideology with a large portion of the money of each of the richest people.

Clearly, c_{vague} is vague, because 'a large portion' admits borderline cases. The vague politicians' game works in the same way as the ambiguous politicians' game: fierce advocates of redistribution would favour c_{high} over c_{vague}, for example, because the latter leaves them uncertain about the amount of redistribution. It is also plausible that politicians would prefer to avoid a commitment as clear as c_{high}, because future contingencies might make it difficult for them to honour this promise. In fact, one could extend the game with a second election, in which the electorate could give their verdict on a politician's time in office; surely, the breaking of promises doesn't do much for a politician's re-election chances, and a precise promise is easier to break than a vague one.

With help from Aragonès and Neeman, we have found a situation in which vagueness has a higher utility than precision. Models of this kind hinge on the fact that the interests of the speaker and the hearer differ: what's good for the politician may be bad for his voters. NLG systems can be faced with similar asymmetries, for example when an artificial doctor decides to keep its predictions vague to avoid being contradicted by the facts; a doctor who says 'These symptoms will soon disappear' is less likely to get complaints than one who says 'These symptoms will have disappeared by midnight'. In both cases, uncertainty about the future is the main reason for being vague. A slightly different type of situation obtains in many advertisements (e.g. 'This hotel offers excellent value for money!'), where a claim is cast in vague language even though the author of the claim knows all the facts. In such cases, the interests of the speaker differ substantially from those of the customer. Examples where vagueness can save money or face are plentiful, in other words. One wonders whether vagueness can also be advantageous in situations where it is one's honest aim to inform an audience as well as one can.[9]

Before we move on to explore this question, one other use of vagueness should be mentioned that also hinges on conflicts of interests. *Laws* are often left vague or ambiguous because lawmakers cannot agree on a particular clause. As a compromise, a vague formulation may be chosen. Consider the contentious issue of abortion, for example. If one political faction favours a 20-week threshold for abortions while another favours a 24-week threshold then the law might refer to 'a reasonable time period', for example. In this way each faction will be able to argue to their supporters that the law says what they would like it to say. Vagueness in contracts (e.g. 'The car must be returned in good condition') can arise in similar ways as in laws. In such instances, the law allows borderline cases, where it is up the discretion of someone other than the disagreeing parties—a judge, for instance—to make the decision.

The famous legal theorist H. L. A. Hart borrowed the concept of *open texture* from the philosopher Friedrich Waismann to describe these phenomena.[10] Legal philosophers differ over how much openness of texture is desirable: vagueness makes the application of the law less predictable, with the possibility of arbitrariness and favouritism. But when people differ over how a concept needs to be defined, it is convenient to appease them by leaving certain issues undecided. Such factors make a certain degree of vagueness almost unavoidable.

Vagueness in the absence of conflict

The question why vagueness is used strategically in situations where the interests of speakers and hearers are essentially aligned was asked perhaps most forcefully by the economist Barton Lipman. Lipman used an airport scenario involving two players. Here is the gist of the story in its simplest form.

> Player 1 asks player 2 to go to the airport to pick up an acquaintance of player 1. Player 1 knows the referent's height with full precision, while player 2 carries a perfect measuring device. There are two other people at the airport, and it is assumed that heights are distributed uniformly between a maximum denoted by 1 and a minimum denoted by 0; this means that every height occurs equally often. The pay-off for both players is 1 if player 2 successfully picks the referent, while it is 0 if the first person she taps on the shoulder turns out to be someone else.

Lipman argues that, under these assumptions, player 1 would be foolish to use vagueness: why say 'he is tall' if you can say 'he is 183.721 cm'? By stating his acquaintance's exact height, player 1 will allow player 2 to identify this person with almost complete certainty, given that the chance of two people having the exact same height is almost nil. Lipman realizes that this strategy would imply that the language is able to distinguish between all possible heights, so he also asks what would happen if only one word were available to player 1, regardless of the person he needs to identify. He proves that, under these assumptions, optimal communication arises if this word (e.g. the word 'tall') is used in accordance with the following rule:

Say 'tall' if height (person) > ½, else say 'not tall'.

This concept does not involve vagueness, because the rule does not allow any borderline cases: everyone is either tall or short. In other words, no rationale for vagueness has been found so far.

Based on examples of this kind, Lipman asks what explains people's abundant use of vagueness. For, on the strength of the airport scenario, vagueness might easily seem 'a world-wide, several-thousand-year efficiency loss'. The difficulty of pinpointing the *rationale* for vagueness becomes even starker if we realize that vague words are invariably among the first words children learn to use and understand.[11] Note that Lipman is not doubting whether vague utterances can be useful, which they evidently can be. He is asking whether vague expressions can be *more* useful than any non-vague expression—or at least equally useful as these. Like us, he is focusing on situations where there is no conflict between the people involved.

Answering Lipman

In what follows, let us consider a number of possible answers to Lipman's puzzle, most of which are also mentioned by Lipman. The upshot will be that some uses of vagueness can be recognized as eminently useful, but that the existence of borderline cases—the hallmark of vagueness—is harder to justify.

First answer: Necessary vagueness

Sometimes I speak vaguely because that is the best I can do. This happens, for example, when all I have is someone else's spoken report: if you tell me there are 'many' students on a given university course and I trust your judgement, then if someone asks me, the best I can do may be to pass on your assessment *verbatim*. Important though this may be, second-hand vagueness does not give us much insight, because it begs the question why I was given vague information in the first place.

It appears[12] that vagueness can also stem from the limitations of human *memory*. Suppose I told you that 324,542 people perished in some cataclysmic earthquake. For a while, you might remember the exact death toll. In the longer term, however, details are likely to be corrupted (if you remember the wrong number) or lost: the next day, you may only recall that the victims numbered in their hundreds of thousands; a year later, you may only remember that there were many. Similarly, I may remember the position of a shooting star with considerable precision after a few seconds, but a minute later I can only point vaguely at some vast section of the sky. Human memory appears to have evolved to economize on memory space, while retaining only the gist of the information presented to it—on a good day, that is.

In search of other situations where vagueness is unavoidable, let us return to the original airport scenario, assuming that the speaker describes the person in question as 'taller than 175 cm'. Lipman saw this as a crisp description of the person, because he assumed it to be based on absolute measurement. But in light of the discussion in Chapter 3, we cannot agree with this entirely, because we have seen that 'taller than 175 cm' is actually *vague*, given the imprecision that affects height measurement. On the other hand, one has to admit that the concept 'taller than 175 cm' is far *less* vague than our everyday word 'tall': surely almost every person is either taller than 175 cm or not; only people whose height is such that it might equally be measured as above or below 175 cm are borderline cases. So although measurement and perception force us to be vague, they do not force us to be as vague as we actually are. In this sense, I believe Lipman to be right: these factors do not explain the extent to which vagueness affects our communication. At the end of this section, when discussing the seventh answer to Lipman, we shall return to the issue of perceptual limitations.

Second answer: Apparent vagueness

We can gain more insight into these matters if we allow ourselves to change Lipman's scenario. In a new, modified airport scenario the speaker knows the heights of all three people at the airport. Suddenly it becomes easier to understand why vagueness can be useful. For suppose your acquaintance happens to be the tallest person of the three. You can then identify him as 'the tall guy'. Arguably, this is safer than citing the person's height in centimetres, because 'the tall guy' does not require the players to make any measurements: *comparisons* between heights can often be made in an instant, and with more confidence than absolute measurements.

To see what I mean, imagine two similarly shaped mansions standing next to each other: one of them is infested by wood-destroying insects and needs to be demolished immediately, while the other will be lovingly restored to its former glory. The first house is 20 metres tall, the other 18. Then evidently, it is safer to tell the demolition company, with whom you are talking over the phone, to destroy 'the tall house' than to destroy 'the house that is 20 m tall', since the latter requires some laborious measurement, whereas the former does not. Much of the time, it is far easier to see that one house is taller than another than to tell which of the two is closer to 20 metres. We discussed cases of this type in the section on vague descriptions, where a generator takes numerical height measurements to produce noun phrases that involve gradable adjectives. In cases such as this, vagueness is only *local*, as we have seen. In the modified airport scenario, for example, the noun phrase as a whole (e.g. 'the tall guy') allows no borderline cases, so one can argue that there is no *global* vagueness here even though no exact threshold is implied.

It is worth remarking here that vague words can be used crisply in many ways. When I say of a gymnastic exercise that it is 'good for young and old', for example, then there is nothing vague about my description: I am using vague words to say that this exercise is good for *everyone*, regardless of age. Or consider the slogan 'bad for bacteria, good for gums', on a toothpaste tube. It would be beside the point to ask 'How bad?' or 'How good?', since the aim of the description is to identify the mechanism through which the toothpaste claims to work: bacteria are killed and this is good for your gums. The removal of vagueness works differently from one context to another. For example, when we say 'His feet were large but his hands were small', we say that essentially his hands were smaller than one might expect given the size of his feet. This is a *comparison* between two sizes: the vagueness in the words 'large' and 'small' is only apparent. Yet another type of apparent vagueness arises in examples like the one we discussed in connection with qualitative reasoning: when we say that 'fast-flowing rivers are deep', we are making a crisp statement about the connection between two phenomena: the speed and the depth of a river. In all these cases, vague terms are used to make a crisp statement.

Although local vagueness constitutes a kind of answer to Lipman, it would be disappointing if it were the only answer one could find, because vagueness, in these cases, can be seen as only apparent.

Third answer: Vagueness as cost reduction

Various authors have suggested that strategic vagueness can arise from a desire to reduce the 'cost' of the utterances involved, as long as this does not cause too large a reduction in their benefits. Broadly speaking, the idea is that vague expressions are easier to produce and interpret than precise ones. This is a plausible angle on vagueness, but it is not easy to substantiate. Let me explain.

First, let us focus on some situations where the goal of the utterance is clear, starting with referential language use. A referential expression such as 'the tall house' is almost certainly easier to produce and interpret than 'the house that is 20 metres tall'. But vagueness is only apparent in such cases, as we have seen, so let us see if the same mechanism might be at work in other situations. Consider a doctor who measures your body temperature as 38.82 °C. By stating that you have 'a high fever' (instead of 38.82 °C) the doctor is pruning away details that are of questionable relevance in the situation at hand. Perhaps this is vagueness as cost reduction. But Lipman might respond by saying that cost reduction is no reason for using language that is vague: language that allows borderline cases, in other words. The doctor could have achieved a similar economy by saying that your temperature is 39 degrees (using the normal rounding conventions), or that it is higher than 38 degrees; either way, she would have reduced information without being vague. Conversely, she could have been vague without achieving cost reduction, for example by saying that the temperature was 'somewhat close to 40'. It is far from clear that vagueness by itself reduces linguistic costs.

Along similar lines, one could argue that vagueness leads to cost reduction in *learning* the concepts involved, because vague words are part of a general mechanism that makes their meaning dependent on the context in which they are used. The size constraints on 'a small elephant', for example, are very different from those on 'a small mouse', but they follow from what it means to be an elephant and what it means to be a mouse. And sure enough, one can see how the *context-dependence* of words such as 'small' makes these concepts easy to learn. But once again, it is difficult to see why this means that they must be vague (i.e. that they allow borderline cases). In fact, most of the mechanisms for context dependence that were discussed in Chapter 6 resulted in a crisp inter-pretation for expressions of the form 'a tall so-and-so', for example by stipulating that this means being a so-and-so who is taller than the average so-and-so.

Fourth answer: Adding an opinion

Frank Veltman (2002) and others have argued that vagueness does not always give us *less* precision, it can also give us *more*. Perhaps the clearest evidence that vague expressions can involve the expression of an opin-ion, or bias, is the fact that they are often used in situations where other functions are difficult to detect. If you tell me you paid 200 pounds for your shoes, I might respond 'That's expensive!' My utterance cannot be designed to tell you what the price is, because you know this. My point in saying this is merely to tell you what I think about it. My utterance is of the form $p > n$, as it were, where p is the price of these shoes, and n the normal price (or the highest acceptable price in my opinion) for shoes of this kind. Since the value of p is a given, the news for you can only lie in n.

An opinion is particularly useful if measurement is problematic. Most of us, who measure coffee in cups or spoonfuls, are none the wiser if a medical information leaflet informs us that a pill against headache contains 25 mg of caffeine. (Is that enough to keep me awake?) The same is true if we hear that our blood pressure is 140/90, for example, until we are told what these figures mean in terms of our health. Expressions such as 'high blood pressure' finesse this problem hand-somely, by telling us there is cause for concern. There is, of course, a fine line to be drawn between commendable simplification and intolerable patronizing; in many situations, the expression of bias is justifiable only if full, unbiased information is provided as well, but that's another matter.

There is, however, a puzzle to be solved here, along much the same lines as before. What has vagueness got to do with opinion or bias? There are a number of options here. The doctor could say that this reading should be considered worrisome (possibly citing the numerical value as well). Why should bias be coupled with vagueness only, given that it is as easy to think of a crisp expression that contains bias as it is to think of a vague expression that does not. An example of the latter is an adjective like 'tall'. An example of the former is the word 'obese', in the sense of having a body mass index of over 30 (see Chapter 3); recall that obesity was defined in this way precisely because this degree of overweight is considered medically worrisome.

Fifth answer: The lack of a good metric

For simplicity, we have pretended that when information is expressed, the starting point is a well-defined chunk of information. Similarly, we have simplified by assuming that the words we use must always corres-pond with well-defined metrics in well-understood ways. In many situ-ations (as Lipman acknowledged), this is far from true. When discussing mathematics and numbers, for example, we saw that the difficulty of a proof is frequently expressed in vague terms, because it would be difficult to measure this objectively. Another kind of example involves the combination of different dimensions. If I call a house large, for example, how exactly do I combine its square footage with its height? (Perhaps that attic, which extends over the entire width of the house but whose height is barely 1 metre, is of no use to me at all!) Many of the things we say are based on much more complex information, whose quantification is immeasurably more subjective than the size of a house.

Suppose someone is happy, or beautiful, or pleasant to be with. If you are like me, you struggle to quantify such things. Introspection—a fallible guide, admittedly—suggests that we are very unlike the computers discussed in previous sections, which read a number off a measurement scale, which is subsequently put into words. All we have is vagueness, it seems, and how this vague information is represented in our heads is a mystery. In the philosopher Mark Sainsbury's colourful words, 'the throbbing centres of our lives appear to be describable only in vague terms'. Sainsbury might be proved wrong, but his words are not easy to contradict given the present state of knowledge. Admittedly, these observations have less relevance for the down-to-earth applications to which NLG has most often been put (weather forecasting, medical reports, etc.), but they are highly relevant to our understanding of vagueness nonetheless.

Sixth answer: Future contingencies

Laws and contracts notoriously contain expressions that can be applied to a concrete situation only after the unclarities in them have been resolved. We have seen that this can sometimes be an attempt to avoid a conflict between people who disagree over some issue (such as, for example, the time frame within which abortions should be allowed; see also Chapter 4). But vagueness can also arise for reasons that have little to do with conflicts between lawmakers. In some cases, vagueness can be seen as a way of *parameterizing* the law and thereby making it applicable to a larger variety of situations, whose details are impossible to foresee. Lipman's examples of this include such phrases as 'taking appropriate care'.

In areas where social norms or technology play a role, this reason for vagueness is particularly obvious. In 1981, for example, the British Parliament passed a law forbidding the public display of *any indecent matter* (except, notably, in art galleries) but it did not specify what was meant by 'indecent'. Another often-invoked example is an injunction against using vehicles in a park. Does this apply to scooters? Bicycles? And how about the newfangled skates that allow you to go very fast? The word 'vehicle' is not well enough defined to tell. The vagueness in these laws means that they stand a good chance of surviving for a long time: the law against indecency, for example, forbids us from doing what is indecent *at a given moment*. The downside is that there are bound to be differences of opinion over what indecency is and how indecent something has to be to be covered by this Act.

Strictly speaking, the fact that such expressions as 'indecent' and 'vehicle' depend on context for their interpretation does not make them vague. (To see this, consider the word 'I', which refers to a different person depending on who is speaking, but always only to just that person, so no vagueness is involved.) They are vague because context affects the interpretation of these words in ways that are impossible to foresee: their precisification depends on who it is that does the precisifying, for example—and what they have had for lunch.

Seventh answer: Vagueness facilitates search

Cast your mind back to the story of the diamond that was stolen from the Chinese imperial palace, back at the start of Chapter 9. The theft was observed by a sole witness, who identified the thief as 'tall'. We have seen time and again that people's understanding of such words as 'tall' is not exactly the same. Rohit Parikh's analysis of the story, in Chapter 7, of Ann's blue topology book told us that such incomplete alignment is not necessarily a disaster, and this is surely true in the case of the stolen diamond. It's better to know that the thief has been described as 'tall' than not to know anything.

But what happens, in the case of the stolen diamond, if the thief is someone the Emperor considers too short to be called tall? The witness says the thief is tall...so if the Emperor follows his classical logician, he searches all the tall eunuchs. If the thief is not amongst them then all bets are off: all the tall eunuchs (more precisely, the ones the emperor considers tall) have been searched, and now it is the turn of the other ones. Lacking a principled strategy, the Emperor has to expect that about half of them will need to be searched before the diamond is found. In this situation, the witness's statement ('the thief is tall') is actually detrimental, because it causes the Emperor to suspect the wrong crowd and lose precious time as a result. The Emperor would have been better off without any description of the thief!

Surely, there is something counter-intuitive about this analysis: after failing to find the thief amongst the tall eunuchs, a smart Emperor should not search the remaining eunuchs blindly, but start with the ones that he *might* have considered tall had he set the threshold for tallness a little lower! This move would have been perfectly natural to the partial logician: after inspecting the eunuchs who are definitely tall, he will search the borderline cases (i.e. the ones in the truth table gap for 'tall'), because these are more likely to be called 'tall' than the ones who are clearly not tall. The fuzzy logican could have been even more efficient by searching the eunuchs in order of their height, starting with the tallest: surely this is what a smart Emperor would have done. The crucial point is that smart moves of this kind are open only to someone who regards 'tall' as vague. If your language contains only crisp concepts, they make no sense: there are tall eunuchs and non-tall ones, and that's it. Only if you understand 'tall' to be vague or gradable can you distinguish between the different people whom you do not consider tall.

At the heart of this seventh answer lies the observation that when perceptual limitations make it difficult for speakers to align the meaning of their words, mismatches are inevitable. When mismatches—between your word 'tall' and mine, or between your word 'blue' and mine—arise then it is advantageous to be flexible. Our understanding of the words in question has to allow us to step into someone else's shoes, so to speak, and ask, 'What if she uses a slightly lower threshold for tall-ness?' Flexibility will allow us to understand other people better and, in cases like the ones discussed here, this will speed up the search the people that others call tall, the books that others call blue, and so on. As I have argued in Chapter 9, it is difficult to see how this flexibility can be achieved within classical logic, and even partial logic allows only a limited amount of it. It appears that only degree theories are able to explain how the information in the witness's statement can be used optimally.[13] I take this to be a promising new angle on Lipman's query.

Why do we speak?

All right: so speaking means balancing the benefits that one gains from an utterance against the efforts associated with it. But whose benefits, and whose efforts? We have focused on situations where the interests of speakers and hearers are essentially aligned, at least to the extent that both would like the speaker to get her message across. But speakers can decide to make more or less of an effort. Consider Rohit Parikh's example again. We are led to assume that Bob eventu-ally managed to find Ann's blue book, but it may have taken him a while. One feels compelled to speak up for Bob! Surely, Ann could have given him a more elaborate explanation, adding that the book was 'somewhere on the bottom shelf', for example, and this might have saved her partner a lot of work.

I realize I am speaking in almost moral terms, asking what a speaker *should* do. The approach to vagueness outlined in the previous section has assumed all along that speakers do what is best for hearers. It might be thought that this is a foregone conclusion: we speak in order to be understood, don't we? A growing body of experimental work, however, suggests that this is not always the case. Surprisingly often, speakers and hearers behave as if they were blissfully unaware of the other person. What makes this particularly interesting for us is that many of the crucial experiments in this area have vague expressions at their heart. Let me sketch a couple of these experiments.

One experiment had two people looking at a set of differently sized candles. The hearer sees three candles, but the speaker sees only the two larger ones, because the smallest candle is obliterated from the speaker's view. The experimenters make an effort to rub the hearer's nose in the fact that the speaker sees only two candles and does not know about the smallest of the three. Yet, when the speaker says 'get the small candle', the hearer will almost consistently choose the smallest of the three even though, rationally, he knows that this one cannot be intended. He behaves, in other words, as if he were unaware of the differences between himself and the other person.

In another experiment, Sid Horton and Boaz Keysar compared speakers in two different situations. In each, the speaker and the hearer are looking at two computer screens. What they see is, for example, a circle that moves slowly from one screen to the other. The speaker has to describe what she sees. In both situations she sees one *other* circle on her screen, of a different size, which allows her to think of the target circle as 'the small circle' or 'the large circle', depending on the size of the other one. In one situation, the hearer sees exactly the same as the speaker. But in the other, the hearer sees only the target circle, *without* another to compare it to. Crucially, the second situation will make expressions like 'the large circle' incomprehensible to the hearer.

In the same way as in the previous experiment, it is made abundantly clear to the speaker that the hearer can see only one circle. Clearly, therefore, if speakers were considerate they would be much more cautious in the informationally asymmetrical condition than in the situation where speaker and hearer see the same things. But as soon as speakers are put under a bit of time pressure, they throw caution to the wind: the experimenters found that there was essentially no difference between the (very substantial) number of degree adjectives used in the two situations. Expressions such as 'the large circle' were used equally often in both situations. In other words, it appeared that speakers did not make *any* allowance for the ignorance of the listener by cutting down on words that would be incomprehensible to their audience: they just blabbered on. Perhaps we shouldn't be surprised: we all gesticulate on the phone.

Experiments of this kind are beginning to suggest a striking reluctance, in most of us, to take other people's views into account.[14] Sure enough, normal adults are able to reason about other people's minds, but in many everyday situations we do not use this ability. This tendency has been likened, somewhat exaggeratedly perhaps, to what might happen if you get a complex espresso machine for your birthday: the machine may be in good working order and yet end up gathering dust. According to Keysar and his colleagues, the same has happened to our ability to reason about other people's knowledge: we are able to do it, but we often cannot be bothered.

This brings us back to the Introduction, where we asked how language and communication should be studied. Should their study be a purely empirical affair, or were philosophers like Russell on the right track when they bemoaned the pitfalls of language? In other words, should linguistic rules be descriptive, or should they tell us how we *should* use language? This age-old debate has obvious implications for everyday life, such as the question of how language should be taught at school. Also, it has a direct parallel in artificial intelligence, where a programmer might wonder what kind of communicative behaviour to build into an NLG program that she is helping to construct.

The theme of language as an instrument for manipulation has featured a number of times in this book. In connection with such words as 'cause', for example, we saw that words can suggest a false clarity that is not justified by the facts on the ground. Something similar held for the false precision inherent in some numerical statements, particularly when the numbers involved are round. In yet other situations, we observed that vague expressions are systematically misunderstood, particularly where small percentages are concerned. We have now come full circle, asking what is more important: description or prescription. Or, casting this in more practical terms, should artificially intelligent systems mimic the way in which people happen to communicate, or should they be designed so as to communicate optimally?

Academically, there is no conflict, because research on communication can take one of two equally legitimate shapes. The first is *sender-oriented*: it attempts to predict what people will naturally say or write in a given situation; this work can be linked to the Turing Test (see Chapter 10). The other kind of research is *receiver-oriented*. It tries to predict which texts work best, for example because they are clearest. There is no conflict between studying the *is* and the *ought* of communication: there is a place for each. But a conflict does arise when practical decisions are made, as when a person decides to say one thing rather than another, or when an artificially intelligent machine produces an utterance. When naturalness and effectiveness fail to go hand in hand, there may be a choice between

- a highly natural utterance which happens to be unclear or misleading, and
- a highly effective utterance which would seldom have been used by human speakers.

As so often, the decision depends on what you want to achieve.

Things to remember

♦ Our discussion of weather reporting suggested that context-dependence is one of the most crucial aspects of vagueness from the perspective of artificial intelligence. If AI programs were able to treat vague concepts (such as good weather, a healthy patient, etc.) in a context-dependent way, then programming errors would become less frequent, and more human-like interaction with computers would become possible.

◆ In a similar vein, AI programs would improve if they were able to play down the importance of arbitrary thresholds (such as the threshold separating light rain from heavy rain), for example when reporting sequences of measurements scattered on either side of a threshold.

◆ *Reference* is a microcosm of human communication, which allows detailed study of the way language works. Our discussion of reference showed that vague boundaries can sometimes become crisp as a result of the context in which they are applied; in such cases vagueness is only apparent. On the other hand, the discussion has also highlighted the complications caused by a highly complex gradable dimension that is always among us, namely salience.

◆ The fact that *pointing* tends to be vague indicates that vagueness is no accidental feature of the languages we speak, but an almost unavoidable feature of communication.

◆ Decision theory and game theory are well placed to shed light on the question why we communicate the way we do. These disciplines highlight the benefits arising from communication, and the costs associated with it. One recent focal point in game theorists' study of language is the question of how vagueness can be useful.

◆ A utility-based perspective can help us understand why vagueness is used in situations where crisp information is unavailable (for example, because no commonly agreed metric is available) or where the interests of hearers and speakers differ sharply (for example, in advertisements); it also allows us to understand why gradable words are used in a context that enforces a crisp interpretation. Words of the latter kind are only apparently vague. Put differently, they are locally vague but globally crisp.

◆ The story of the diamond stolen from the imperial palace suggests that, when mismatches between people's understanding of words occur, vagueness can facilitate the search for the intended referent of a description (e.g. 'the tall man', 'the blue book').

12

The Expulsion from Boole's Paradise

What have we learned? Long ago, Aristotle wrote that 'a well-schooled man is one who searches for that degree of precision in each kind of study which the nature of the subject at hand admits' (*Ethics* I. 3). These lines can still send a blush to the face of anyone spending time with formal models of vagueness: perhaps by insisting on exact models of things that are, by definition, not exact, we are trying to achieve the impossible. Have we been chasing the wind?

Although I do not claim to have all the answers, the reader is entitled to a general conclusion—or at least something like it. I shall pay this debt by addressing the questions that were asked in the Prologue. I rearrange them slightly in the light of the previous chapters.

Earlier questions revisited

Are all vague concepts essentially alike?

Perhaps the main difference between vague concepts that has come to light is that vagueness itself is a matter of degree: some concepts are vaguer than others. The word 'tall', for example, tends to involve a large range of borderline cases. By comparison, the word 'poisonous' is less vague, and an expression such as 'larger than 176 cm' even less vague than that, because it does not have any borderline cases except in the vicinity of 176 cm. This is why a report that 40 per cent of Britons are tall (without explanation of how the word is used) would be greeted with more scepticism than a report that 40 per cent of people are taller than 176 cm, or that 40 per cent of food colouring agents are poisonous.

From this relativistic stance, the vagueness of a word may be likened to the smoothness of a physical surface or the straightness of a line. Some lines are clearly bent whereas others appear to be straight; but under a magnifying glass even the straightest of lines will show up imperfections. Similarly, apparently crisp distinctions have a tendency to turn vague when examined closely. Perhaps the real thing—a perfectly straight line, a perfectly crisp concept—exists in our imagination only.

A few other differences between vague concepts are worth reiterating. First, some concepts are associated with well-understood and commonly agreed measurement scales, but others are not: to establish objectively how intelligent a person is (see Chapter 3), let alone how friendly, would be problematic. Secondly, some vague concepts which we do know how to measure—such as obesity and poverty—have been 'sanitized' by means of artificially crisp boundaries that stop the concept from being vague in the first place (apart from the small residual vagueness associated with measurements of height and weight). Thirdly and lastly, some vague concepts relate to the world around us in much more complex ways than such concepts as 'large' or 'tall'. In Chapter 7, for example, we discussed the case of colour perception, which makes some colour patches physically very unlike each other seem equal to us. Other differences between vague concepts have been noted, but these are the ones that stand out.

Why do we use vagueness so frequently?

The previous chapter has discussed a variety of situations in which speakers tend to make use of vague words. In some of these situations, vagueness is only apparent. When we say that vegetables are good for 'young and old people alike', for example, vagueness does not play a role: we are saying only that vegetables are good for us, regardless of age. Similarly, when we say, 'When there is much water in the bathtub, a lot of water will flow through the drain', we are expressing a correlation between the amount of water in the tub and the amount of water flowing out of it in any unit of time: an increase in one means an increase in the other. Vagueness does not play a role. A degree adjective does not necessarily signify vagueness, in other words.

This does not mean that genuine vagueness is rare or unimportant, or that it is just a leftover from the primitive days when our inarticulate ancestors roamed the earth, as some would have it. One version of the theory that vagueness is essentially a flaw can be found in the amusing book by Gary Marcus, *Kluge*. As a clue to his views, he offers the following explanation of the word 'kluge': 'noun, pronounced / klooj/ (engineering): a solution that is clumsy or inelegant yet surprisingly effective'. Vague language use, for Marcus, is typical for the kluge-like minds that we have inherited from our primitive ancestors.

We are capable of much greater precision than our ancestors, and it is only to be expected that our language reflects some limitations that no longer exist. Yet, vagueness is often unavoidable, as we have seen time and again. The history of the metric system, for example, shows that many types of measurement cannot be pinned down with absolute precision. As a result, concepts such as '175 cm tall' and 'taller than 175 cm' are, strictly speaking, vague, because there are objects regarding which even the most sophisticated measurements available cannot establish with certainty whether these concepts apply. In many other situations we do not even possess the beginnings of an objective metric that would allow us to avoid vagueness. How, for example, do we measure the beauty of a sunset, the wisdom of a person, or even the strength of a football team? It is in view of these issues that vagueness was likened to original sin: a stain that can be diminished but never removed.

The vagueness of many everyday concepts, appears to be caused by the lack of a well-understood metric. A species-denoting term such as 'tiger' has been left vague not because its vagueness involves a cost reduction or an opportunity to express an opinion; it is vague because speakers have not settled on a common criterion. Interbreeding was an attempt by biologists to find such a criterion, but it proved to be flawed, as we saw in Chapter 2. Analogous considerations apply to the *identity* of an object, which played a crucial role in the court case involving the vintage car Old Number One (Chapter 4), but even in simpler matters, such as the question when to regard two volumes as the same book. Analogous to scientists' efforts to make obesity crisp, one might design a crisp criterion for deciding the identity of a car or a book, but the result would probably be arbitrary and short-lived. The fact that two books can be very similar is not enough in itself to force us to define their identity vaguely: it is because books are also such complex and multi-faceted things that we find it difficult to define their identity sharply. The same goes for cars and, even more evidently, for people (Chapter 4).

Moreover, even when vagueness is avoidable, it can be useful, and this is where game-theoretical analyses have made a significant contribution to our understanding. Ultimately, such analyses should help us to understand what benefits arise from an utterance, and at what cost. Analyses of this kind turned out to be particularly useful when there are important discrepancies between speakers and hearers, for example because the speaker wants to diminish the likelihood of being contradicted. One example where this was true involved a politician who wanted to keep his options open; the same reasoning applies when a manufacturer advertises his products: vague claims are less easily challenged in court than crisp ones, and this may well explain their prevalence in advertisements.

Another type of discrepancy arises when different people understand a concept differently: our story of the stolen diamond illustrated how unwise it would be to insist on our own understanding of a concept such as 'tall' if we want to understand what someone else is trying to tell us: it is better to be flexible and understand 'tall' as admitting borderline cases (or, better still, degrees), because this will allow us to use smarter search strategies. It is useful, in other words, to understand 'tall' as a vague concept.

Why are vague expressions context-dependent?

Vague expressions mean different things in different contexts. The reason, I think, is that context-dependency helps to make vague adjectives and adverbs widely applicable. We *could* have used different adjectives for a small spider, a small elephant, and a small planet—a bit like naturalists who use a baffling number of different words for different kinds of herds (flock, pack, flight, drove, gang, gaggle, bevy, etc.) just because a different kind of animal is involved. This would have been much harder to learn. It would be easier to reuse the same words over and over again. But this overloading would have meant that the threshold for 'small' must vary depending on the kind of animal, or else the word would be inapplicable to elephants. Vague words, in other words, would have to be context-dependent...as, in fact, they are.

Once again, differences between vague expressions complicate the picture, because some words are more context-dependent than others. At one extreme, words such as 'large' and 'small' are context-dependency personified: choose any number, and you can think of a context in which that number is regarded as small, and another in which it is regarded as large. The same is *not* true, for example, for 'being 1 metre tall'. It is true that the word 'metre' can be interpreted in different ways: in ordinary conversation, its precision is at most a few centimetres; in construction it is more like a few millimetres; in a physics laboratory, its precision may be close to the maximum precision achievable at the time. Yet, a metre is always *roughly* the same distance. Certain things are just too small, or too large, to ever count as a metre. Concepts such as 'metre', which were designed in an attempt to be precise, are context-dependent only in the amount of precision with which they are used. They are constant in all other respects.

What do vague expressions mean?

When vague words are used, paradox looms. In the light of our earlier explorations, we rephrase our original question as 'What is the best logical model of vagueness?' Our discussion in the middle part of the book suggests a nuanced answer. A scientific model is almost invariably a simplification of reality, constructed for a purpose. A sensible person will choose the simplest model that achieves her goals. Consider a down-to-earth example: traffic police use simple graphics to depict the flow of traffic in a city, using the width of an arrow to depict the amount of traffic on a road, for example. We do not criticize these models for failing to distinguish between a Mercedes and a Ford car: this difference is irrelevant to the resolution of traffic jams. (A director in the car industry who wants to monitor the use of the different types of cars will need a very different model.) The quality of a model, in other words, should be judged in relation to a purpose. With this in mind, let's categorize the different models of vagueness.

Classical logic. These are the simplest models on offer. Exaggerating slightly, classical approaches can be construed as saying 'Don't bother modelling the existence of borderline cases. Boolean (i.e. two-valued) dichotomies aren't that far off the mark.' True enough, classical models can build useful applications in artificial intelligence, as we have seen; in many cases, more sophistication would be useless. It would, however, be wrong in my opinion to defend classical models by arguing that words such as 'tall' are only *apparently* vague, as a result of language users' ignorance concerning their crisp boundaries (see the discussion of epistemicism, or vagueness as ignorance, in Chapter 7).

Partial models with supervaluations. These models form an elegant compromise between logical tradition and empirical nuance. Borderline cases are acknowledged, but that's the only concession to vagueness that these models make. The result is a relatively simple theory that

282

shares many of the core features of classical logic, such as the laws of non-contradiction and of the excluded middle. It does not, however, do justice to the intuition that small differences do not matter, as expressed by Kamp's tolerance principle. Also, it cannot explain fully how vague concepts are used in search situations, as we saw in our discussion of the story of the stolen diamond.

Context-based models. Although they are further removed from classical logic, some context-based models display considerable elegance. They shed a revealing light on our awareness of the consequences of judgements that involve vague words and, if certain adaptations are made, on people's inability to notice gradual change. They are, however, kept artificially simple by disregarding the *variability* of human judgements.

Models based on introspection. The model based on margin of error (MoE) and introspective agents formalizes the tolerance principle in a manner that takes into account that an individual can make different estimations regarding the same perceptual phenomenon on different occasions. Matters are kept manageable by pretending that an MoE is always crisp and non-probabilistic, and by disregarding the context effects on which the previous set of models focused.

Degree models. Degree theories may lack the austere elegance of some of the other models, but they have much to offer in exchange. For instance, they allow us to say that small differences do not matter *much*; this graded version of tolerance allows a subtle take on the sorites paradox. The well-known shortcomings of fuzzy logic in this connection (stemming from its truth functionality) can be resolved if a *probabilistic* approach to truth degrees is chosen, as defended in Chapter 9. Degree theories also give us a promising starting point for understanding how people respond to such expressions as 'the thief is tall'. It appears that theoreticians' resistance to degree theories may be partly irrational, deriving from Dawkins's 'tyranny of the discontinuous mind'.[1]

Since each of these models highlights different aspects of vagueness, it would be natural to explore combinations: degree models that are context-aware, for example. By combining insights from different corners, such hybrid models should allow us to gain an even better understanding of vagueness.

Coping with vagueness

Traditional theories of language and communication differ from theories in physics and engineering by relying on models of the world in which there is no natural place for degrees. Symbolic logic looms large in these areas, and the logical theories employed in theories of communication tend to use two truth values only; they are essentially Boolean, in other words. If one decides to adopt a view of communication in which *degrees* take pride of place, and where there is more to the fidelity of a statement than truth or falsehood, we are in a better position to understand how vagueness works and why it plays such an important role in communication. On the other hand, life will become harder: we get expelled from George Boole's paradise (see Fig. 20)! I will sketch some of the dangers that lurk in the inhospitable world outside the Boolean gates. Since I do not have a remedy against them, I will be brief.

Truthfulness and lying

If degrees of truth are adopted, the question whether a statement is truthful is no longer clear-cut. If a real-estate advert describes the puny garden of your house as 'sizeable', for example, is this a lie or a mere exaggeration? Other, well-publicized examples are easy to find, as when President Clinton denied having sexual relations with one of his interns, whereas later evidence suggested that they did share an affectionate moment or two. More importantly, if a representative of the tobacco industry asserted that her employer, a few decades ago, did not know of a link between smoking and cancer, was she lying or merely clinging to a point of view that was starting to be untenable? If you like your morality simple, degree theories are no help.

FIG. 20 *Expulsion from Paradise*, by Albrecht Dürer

Falsification of theories

The problem is not confined to simple statements of fact. Philosophers of science are interested in lawlike sentences, including simple ones such as 'All ravens are black'. How can such a claim be assessed? A classic answer, deriving from the work of the philosopher Karl Popper, is this.

285

To prove this generalization correct, one would have to check every single raven, which is not feasible. On the other hand, it is possible to prove the generalization *wrong*: find one white raven and you will have falsified the claim. Popper's view has been qualified by the realization that, in actual practice, one counter-example is seldom enough: maybe the raven that you found was only painted white by some prankster, or maybe your cheap binoculars distorted its colour. Yet essentially, Popper's account is still the backbone of many people's view of science. It is also many people's favourite view of rational thought more generally: surely, a bona fide idea must, at least in principle, be testable. Sometimes this can be done by proving the idea correct (that is, by verifying it); more often, it can only be proved to be incorrect (that is, by falsifying it). If an idea cannot be tested in either of these two ways then, surely, it can only be dogma, so this standard bit of reasoning goes.

But there is a problem, and it is called vagueness. No raven is pure black, since this would mean that its feathers reflect no light at all. If feathers happen to come in essentially two different extremes (a dark extreme and a much lighter one) then there is no problem, but if there is something like a continuum of feather shades then the biologist needs to sharpen her claim before it can be assessed. The surprise here is not about science. It is, after all, well known that science proceeds by attempting to make its concepts (black, metre, obese—see Chapter 3) precise. What is puzzling are the consequences for everyday conversation. We have seen that vagueness permeates almost everything we think and say. If it is impossible to test the ideas we habitually exchange in such areas as politics, business, and everyday life, because verification and falsification fail to apply, then should all these ideas be written off as mere dogma?

Belief revision

Another way to see the problem is to ask what happens when you hear someone asserting a claim. Simplifying a little, the standard answer goes like this: *you decide whether or not to accept the claim as true; if you do, then you discard all conceivable situations where the claim does not hold.* But now suppose the sentence contains vague information, as when someone says 'The thief is tall', to help you identify a criminal. It is unclear who can be discarded as a potential referent: although you would be well advised to start regarding tall people with suspicion, the thief could in principle be anyone, as long as they are not very short. In situations like this, it is unclear how your beliefs ought to be revised in the light of the new information. In the presence of vagueness, the simple idea of discarding situations has lost its appeal as a model of belief revision, and any alternative theory is bound to be more complex.

Combinations of vagueness and probability

Vagueness often does not come alone. Consider a doctor predicting the course of an illness, saying 'the symptoms should disappear in a month or so'. There is vagueness here of course, but uncertainty as well, as expressed by the word 'should'. It is natural to model uncertainty using probability. But what happens if vagueness too is modelled using probability, as we have proposed? How do the two kinds of probability mix? Recent work in logic is starting to investigate combinations of fuzzy logic and probability, but combinations of two kinds of probability pose challenges of a new kind, which are well beyond the scope of this book.

The role of context

Similar things can be said about context. In Chapter 8, we argued that the insights afforded by various context-based theories of vagueness are undeniable. But if, as we have argued in the previous chapter, vagueness should be viewed along probabilistic lines, as proposed above, then how should these two insights be incorporated? How, in other words, can *context* coexist with *degrees*? The same question can be asked about introspection and the role of margin of error, each of which we might like to take with us on our flight from Boole's paradise. The holy grail, I believe, is a theory of vagueness that allows one to think about truth in gradable terms while, on the other hand, taking into account that assessments of truth are made in a context in which other assessments may already have been made, and in which the observer has an awareness of her own limitations. The truth degree of a statement such as 'Mr 183 is a short basketball player', for example, should depend on the norms, the observations, and the margin of error of the speaker. Additionally, they should depend on the heights of other baseball players, and particularly on height judgements that have previously been made. How these different perspectives dovetail is as yet unclear. There is an obvious need here for more empirical research, and for new theoretical models that are able to explain the results of this research.

Summing up, it is often justified to simplify by thinking about the word in crisp, Boolean terms. This is what we do when we describe the world in terms of *objects* (see Chapter 4) and *species* (see Chapter 2), and when we insist that a given property must always be either true or false of an object, never something in between. But there are drawbacks to these simplifications because, as we have seen in the first part of this book, the *false clarity* inherent in this approach can cause misinformation.

We can now see the flipside of this coin. When these Boolean simplifications are abandoned and degrees are taken seriously, time-honoured ideas become untenable. When this happens, it becomes necessary to invent new ways of thinking about truth, meaning, and communication. Paradise it might not be, but life will be the more interesting for it.

Epilogue
GUARANTEED CORRECT

They must have been hanging there for years: two old barometers, on opposite sides of the window in the antique shop. They are gathering dust.

One of them indicates 28.6, the other 29.9. I guess they are using an old scale, now obsolete. Luckily some words are printed alongside the figures. The reading 28.6 corresponds with the designation MUCH RAIN, 29.9 aligns with FAIR. The figures don't mean much to me but the words suggest that at least one of these instruments may have known better times.

Come to think of it, the weather is perfect, so maybe *both* of them are in need of repair. They were made almost two hundred years ago. Their mechanism, driven by contractions and expansions of mercury, and transmitted by a thin canvas thread that turns a wheel, must have always been affected by lots of unwelcome factors. Temperature, humidity, and so on. Now, two centuries later, rust and wear must have overwhelmed the one thing they were supposed to measure: barometric pressure. But they're still things of beauty. Pity they're so expensive...

My interest has not escaped the antiquarian. Harrassed by his offspring on this Saturday morning, and barked at by a scruffy shepherd dog, he still finds time to offer me a chair and a coffee. Why am I interested in barometers?

Reluctantly, I tell him about my work. I explain how handsomely the combination of figures and words on his barometers illustrates the virtues,

FIG. 21 Edwardian 'banjo' barometer
with numbers and vague words

and the pitfalls, of communicating through such vague expressions as 'much rain'. And how every measurement has limited accuracy. These machines, evidently, are a case in point. A very strong case, if he sees what I mean. (Secretly, I pat myself on the back. Isn't this a clever way to talk down the price of the barometers?)

Here he interrupts me. This cannot be. Yes, one of the two barometers may be faulty. But he can fix that. Really, this is no problem at all. They'll be absolutely perfect! How can I doubt him?

I try to explain that I did not mean to criticize his merchandise. Far from it. Yes, they indicate different values, but what do you expect? All measurement is imperfect. It's the human condition.

Now he's really getting fired up. *Every* measurement imperfect? Do I deny the fruits of technology? Look around me: his watch, his computer, my own glasses even: it's all hi-tech. And 'hi' means 'high precision' if you ask him.

The doorbell is ringing. Another customer demands entry. Nothing is easy for the shopkeeper today, but after a while he is back in his seat, chatting with his two customers. The new arrival seems to know the antiquarian rather well. For me, he is a godsend. An engineer, Norwegian, just returning from his oil platform in the North Sea, sojourning here in Aberdeen. The man measures things for a living, would you believe it. Drilling components, and tiny ones at that. Everything has to fit. You make a nut that's just a little bit off, it won't fit its bolt properly.

The shopkeeper tells me all this, hoping that some of the technical prowess of his acquaintance will rub off on him. The engineer nods to everything he says, modest under his praise. Yes, his measurements are quite precise and accurate. High precision machining, well yes, that's what people used to call it a few decades ago. What was high then has become ordinary today.

The antiquarian winks at me meaningfully. He thinks he has won the day. I query the engineer: *how* accurate are his measurements, on average? About one thousandth of a millimetre. Yes, that is incredibly accurate. But is it perfect? By no means, says the engineer. Yes, the nuts will fit the bolts, but not as snugly as one would like. Fluids sometimes creep into the space between their spirals, causing trouble. What's worse, some of his competitors are starting to crack this problem. He's having to raise his game, or else.

While chatting, the Norwegian and I have stood up from our chairs and moved closer to the barometers. I'm casting a triumphant glance in the direction of the shopkeeper, who is staying in his chair, a despondent look on his face. Why is the Norwegian not supporting him? Has he lost the argument?

The Norwegian is starting to warm to the instruments. He strokes the polished wood. I point him to some elegant details. The best one is the text at the foot of the mercury column, where its says, in beautifully calligraphed letters, 'guaranteed correct'. He joins in my haughty laugh. What measurement tool can be guaranteed to be correct? The other one is also guaranteed I suppose? The shopkeeper is not amused.

I've gained the upper hand and start plotting my next move. Shall I offer half the asking price? Surely I must have a chance now...Suddenly the shopkeeper appears next to the Norwegian. So what do you think, Gunnar? Reasonable price, hey? I didn't know you were interested in barometers. I've had these for years, you know.

He carries on with a smile: True, they may not always be 100 per cent accurate, but who cares? Our learned friend has just taught me that no measurement tool can be perfect, so what do you expect? You can view that inscription 'guaranteed correct' as an ironic comment on the art of measurement. And, come to think of it, they'd look fabulous in your hallway!

When I close the shop door behind me, the two of them have started boxing the barometers.

NOTES

Prologue

1 The best-known illustration of informal rigour can be found in Kreisel (1987), which focuses on the thesis that computability is accurately characterized by, amongst others, the mathematical concept of a Turing machine. Because computability is an informal concept, this thesis cannot be strictly proven, but Kreisel discusses reasons for believing the thesis to be true nonetheless.

Chapter 1

1 Questionnaires on subjective safety have also been applied to the inhabitants of war-torn Iraq, and the figures there are remarkably similar to the ones recorded for British students. Research commissioned by the BBC and USA Today found that, asked whether they felt safe, 40% of respondents said yes in 2004 (the year after the toppling of Saddam Hussein), 63% in 2005 (when democratic elections had just been held), and 26% in 2007. The BBC headed their report of the 2007 questionnaire 'Pessimism growing among Iraqis'. Alternative explanations are possible. It is conceivable that Iraqis had changed their perception of what it means to be safe, perhaps as a result of exposure to the safer world outside their country. ('Safe', in 2007—but not in 2004—might have meant 'as safe as Geneva', according to that optimistic theory.) Words such as 'safe' are sensitive to the context in which they are used. This phenomenon will be explored in the second part of this book.

2 The figures about mass disturbances are from Pan (2008: 270–1), 'Unrest was on the rise, with police struggling to contain an average of two hundred "mass disturbances" every day in 2004, some of them violent clashes.' Pan's book is an excellent discussion of social issues in today's China.

3 Degree of toxicity can be measured in many different ways. One type of definition involves the notion of a minimal lethal dose, expressed in milligrams per kilogram of a victim's body weight. A much-used version of this concept is LD50, which expresses the lethal dose that kills 50% of a particular type of animal (e.g. rats). Some metrics that are applied to medicines compare harmful effects to desired ones. The *margin of safety* of a medicine, for example, divides the dose that kills 1% of animals by the dose that is therapeutically effective in 99%: the higher this fraction, the safer the medicine (Timbrell 2005).

4 The assertion that people cannot hear the difference between certain sounds is not contradicted by the observation that people can hear differences without being fully aware of it. It is, for example, common for subjects to say they cannot hear the difference between two well-chosen pitch heights while, on the other hand, overwhelmingly answering correctly when asked which of the two is higher (i.e. when they are 'forced' to choose). The effect disappears when the difference in height becomes too small.

Chapter 2

1 The name 'ring species' appears to stem from Stebbins (1949). (See also Wake 1997.) The name was probably influenced by the fact that the habitat of the California Ensatina subspecies has a roughly circular shape. Our analysis suggests that this shape is accidental, and that the expression 'string species' might have been more appropriate. I was first introduced to ring species through Dawkins (2004). Ensatina salamanders are discussed in Pallen (2009), in the chapter 'The evidence for evolution'. Also instructive is the subsequent discussion of speciation as 'the production of a novel population of individuals that can no longer breed with members of the progenitor community or with any other population' (ibid. 89). For more on species and speciation, see Ptacek and Hankison (2009).

2 Dawkins uses the terms 'continuous' and 'discontinous' informally. It appears that, to him, a concept is discontinuous if it must always either apply or fail to apply; this makes 'discontinuous' synonymous with 'crisp'. 'Continuous', for Dawkins, means what we call 'vague'. The examples discussed in this chapter demonstrate that a situation can be essentially *discrete* while still giving rise to vague concepts. The mathematical concept of continuity will be discussed at the end of Chapter 7.

294

3 Plato's theory of ideas, which was attacked by biologists such as Ernst Mayr, found its most famous form in his parable of the cave. The bottom line of the parable is that the things around us are only feeble shadows of a world that is much more real than they themselves. According to this view, individual dogs such as Fido and Towser are only half real: the real deal is the immutable concept 'dog' which they both mirror but imperfectly. Latter-day Platonists do not hold such extreme views, of course.

4 Relations can fail to be transitive for more than one reason. Given the shape of the earth, for example, the relation 'is to the east of' is non-transitive for a different kind of reason from the ones that have occupied us in this chapter: Tokyo is to the east of London, San Francisco is to the east of Tokyo, yet San Francisco is normally thought of as west (not east) of London. This is not because of vagueness, but because of circularity.

Chapter 3

1 The history of distance measurement is a standard item in many engineering texts. Nothing hinges on the metric system per se: from 1959 onwards, the Anglo-Saxon system of yards, feet, and inches has been linked to the metric system, so changes in the metric system automatically trigger analogous changes in the Anglo-Saxon system. My discussion of body weights owes a debt to the lucid discussion of animals' weights in Dawkins (2004). A remarkably (given its age) insightful discussion of vagueness in measurement can be found in Russell (1948); his discussion on pp. 274–83 hints at the sorites paradox.

2 How difficult it was for physicists to escape from ostensibly vague definitions of the *boiling point* of water is nicely documented in Chang (2007), who describes how early physicists used qualitatively defined notions of boiling based on observation by the naked eye. The result was a variety of different degrees of boiling, each of which was only loosely defined. One example is a proposal made by De Luc in 1772, who distinguished six phases: *common boiling* (the highest phase of boiling); *hissing*; *bumping*; *explosion*; *fast evaporation*; and *bubbling* (the lowest phase). Common boiling, for example, was defined vaguely as 'numerous bubbles of vapour rise up through the surface at a steady rate' (Chang 2007: 20). Modern definitions of boiling focus on the amount of vapour

produced, e.g. 'boiling' is the temperature at which the vapour pressure of the liquid equals the environmental pressure surrounding it.

3 There is an extensive literature on the assessment of people's weights (see e.g. World Health Organization 2000). It is particularly interesting to see how definitions of obesity are taking shape in areas where agreement has not yet been reached, for example where the people measured are children (Cole et al. 2000).

4 Concepts of absolute and relative poverty are discussed in many textbooks on economics, including Ray (1998), which focuses on low-income economies. An interesting newspaper article analysing the increasing wage gap in the UK is Thorniley (2007).

5 For a discussion of racial bias in IQ tests, as a result of which the IQ of speakers of black vernacular English may have been systematically underestimated, see Labov (1972). For the case against test bias, see Jensen (1979). For a discussion of the Flynn effect, see Neisser (1997).

6 The dialogue fragment presented here focuses on the cyclicity of comparisons. See Lehrer and Wagner (1985), Fishburn (1986), Logvinenko et al. (2003), and the examples given in these papers.

7 Versions of Putnam's influential ideas on word meaning can be found in many parts of his extensive work. A good start is the essay 'The meaning of meaning' in Putnam (1975–83), ii. *Mind, Language and Reality*.

Chapter 4

1 For a monologue-size treatment of the issues discussed in this chapter, see Parsons (2000). For the case of the racing car, see Forbes (2007).

2 Robert Musil's famous novel *Der Mann ohne Eigenschaften* was written mostly in the 1920s and 1930s, and never finished. The first 'complete' German edition of this classic novel appeared in 1987. The quotation that I used at the start of the book is from the English translation of 1995 by Sophie Wilkins, entitled *The Man without Qualities*.

3 There is some debate about whether vagueness can reside in the world itself, rather than in our speaking about it. Some authors believe that it is in the nature of a cloud or a mountain, for example, to be vague (Morreau 2002). I find this position difficult to understand but I concede that the world often 'invites'

296

vagueness. Pinning down precisely what a name such as 'Mont Blanc' refers to would be difficult, but this does not mean that Mont Blanc itself is vague. For more discussion, see Chapters 7 (section on 'Continuity and vagueness') and 11 (section named 'Answering Lipman').

Chapter 5

1 A delightful and instructive book about the art of approximation is Weinstein and Adam (2008); it focuses on particularly difficult approximation problems, usually with the relatively modest aim of arriving at an answer that lies within a factor of 10 of the correct answer. (For example, if the correct answer is 100 then approximations between 10 and 1,000 are considered acceptable.) Such problems are sometimes known as *Fermi problems*, after the physicist Enrico Fermi who identified and answered many of them.

2 The mathematics of infinity is full of intriguing surprises. If you take an infinite set and duplicate all its elements, then the resulting set—which one might have thought to have doubled in size—has the same size as the original, for example. Yet, infinite sets do come in different sizes. For a brief introduction to the mathematics of infinity, see Rosen (2007). For more detail, see van Dalen et al. (1978); for a thorough treatise, see Drake (1974).

3 The text of this chapter speaks loosely of numbers that can be represented 'efficiently' on a computer. By 'efficient', we mean representation using one *word* (i.e. fixed-size string of bits). At the time of writing, maxint equals 2,147,483,647 on many smaller computers, that is 2^{31}. Larger computers go as far as maxint $= 2^{63}$. Methods for representing much larger numbers do exist, but they are subject to limitations of their own. Numbers higher than maxint can often be represented using vectors of numbers, in which case it is the memory size of the computer that determines what numbers can be represented. The argument in the text can be reproduced precisely regarding this notion of a 'really' maximal integer. Computational techniques for representing certain infinite objects do exist in some computer languages (such as Haskell), but they depend on clever *lazy evaluation* techniques that represent only as much of an infinite object as is required for a particular calculation.

4 Krifka's (2002) principle 'be vague' may sound surprising, but see Chapter 11 for a discussion of reasons for being vague. Krifka's story about English is

complicated because the language contains remnants of number systems preceding the decimal one, based on multiples of twelve. Owing to these remnants, English has short words for numbers that would otherwise have required longer expressions. In other words, Krifka predicts that expressions such as 'He bought a dozen sweets' are less precise than 'He bought twelve sweets'. The claim that round numbers are interpreted vaguely was recently confirmed in an empirical study based on annotated language corpora (Hripcsak et al. 2009).

5 Computational complexity is a staple of theoretical computer science. For a beautifully informal introduction see Harel (1989), particularly Chapter 6 on the efficiency of algorithms ('Getting it done cheaply').

6 Statistical significance is explained in dozens of textbooks and in countless courses on the World Wide Web. One often-used textbook full of worked examples and exercises is Johnson and Kuby (2000). A nice discussion of some of the pitfalls of inferential statistics is Goldacre (2009).

7 Descriptive statistics is an area of 'boring certainties' only when errors are disregarded. How important this qualification is is explained at length in Blastland and Dilnot (2008), who show that official data on just about everything that concerns people is riddled with stark errors.

8 The reasoning about larger samples is analogous to the reasoning about samples of one. To compute the figures in our example involving American and Russian heights, one can use a Z test, which would divide the difference between the American mean (175 cm) and the sample mean (177 cm) by the result of dividing the standard deviation (2 cm) by the square root of the sample size. The resulting Z score tells you how 'extreme' the average of your sample size is (i.e. how far removed from the hypothesized average), and this is easily converted into a probability, for example by using a look-up table.

Chapter 6

1 Linguists do not just *study* vagueness: the concept lies at the heart of their discipline. One standard view of the division of labour between linguistics and phonetics (the science that focuses on the sound of language) holds that linguistics is all about *crisp* aspects of communication, whereas phonetics is about *gradable* issues. Consider the words 'bat' and 'bad', for example, as they are spoken. The plosive sound at the end of 'bat' is a bit more abrupt than in

298

'bad'. It would be easy to produce sounds that are somewhat in between, so they inhabit a gradable scale. Linguists, however, regard 'bat' and 'bad' as entirely different words. Sound, with its shades and nuances, is for phoneticians. This division of labour, whereby linguists focus on what is crisp and phoneticians on what is continuous, is starting to be called into question because of the rise of statistical methods in linguistics.

2 For a brief introduction to the linguistic ideas of Noam Chomsky, see Hornstein (1998). Although Chomsky's linguistic theories have evolved rapidly in later years, Chomsky (1965) is still a very useful introduction. The standard reference for Richard Montague's works is the collection Montague (1974). For an elaborate and accessible introduction to his ideas, see Gamut (1991), which contains a section on Chomsky's work as well.

3 The idea of constructing formal representations of the meaning of a sentence in bottom-up fashion can be found in the writings of the philosophers Gottlob Frege and Rudolf Carnap. Montague showed that this idea is even viable for certain cases (the *intensional* constructions mentioned in the body of this chapter) which initially appeared to resist this approach. Implemented language-interpreting systems did not have to wait for Montague because they tend to disregard intensionality (e.g. Woods 1968). We focus on Montague because he gave compositionality a proper mathematical basis, making the idea more generally applicable and academically respectable.

4 This chapter uses the TENDUM system, developed by Harry Bunt and other Dutch computer linguists in the 1980s, as an example of a question-answering system. Other systems in the same tradition are PHLIQA (Bronnenberg et al. 1980) and SPICOS (e.g. de Vet 1988). The most meticulously worked-out representative of this line of work is the ROSETTA Machine Translation system (Rosetta 1994), which reinterprets Montague's ideas in a deftly simplified way. Perhaps the most linguistically savvy Montague-style question-answering system is the Core Language Engine, developed at SRI Cambridge around 1990 (Alshawi 1992). For a mathematical exploration of compositionality, see Janssen (1997). For a discussion of the problems caused by ambiguity to question-answering systems, see Alshawi (1992) and van Deemter (1998). Some of the ideas on the context-dependency of degree adjectives are too old to trace with certainty, but a good early source is Bartsch and Vennemann (1983).

NOTES TO PAGES 104–119

5 Our discussion of the notion of an adjective is based on Quirk et al. (1972), sects. 5.4 and following.

6 The view that the meaning of vague expressions can, to a large extent, be settled by fiat is starting to gain wide acceptance. Proponents include Kyburg and Morreau (2000), Kennedy (1999), van Deemter (2006), and van der Sluis and Mellish (2009). Kyburg and Morreau's version of this view is of particular interest because it tracks how the meaning of expressions such as 'small' can change as a discourse progresses—'dynamically' as linguists say. It also underlies the computational implementation in a small but interesting language understanding system (DeVault and Stone 2004).

7 For a thorough discussion of the linguistics of vagueness, ambiguity, and indeterminacy, see Pinkal (1995). For a collection of papers focusing on the question how humans and machines can deal with ambiguous input, see van Deemter and Peters 1996.

8 Two exceptions to the rule that vagueness arises from individual words are worth mentioning. The first exception arises when a noun is put in the plural to designate a rough generalization: 'Peasants are poor,' for example, expresses a vague generalization, which we do not expect to hold absolutely. This vagueness stems not from the word 'peasant' or the word 'poor', but from the way in which the words are combined. The second exception concerns the progression of time. In a narrative where a succession of events is related, the fact that one event follows close on the heels of another can be expressed without words: 'He came, he saw, he conquered' relates three events; the claim that these happened in close succession cannot be ascribed to any of the words in the sentence, but only to the way in which they are combined. The narrative does not say how much time passed between the different events, which is where vagueness comes in.

9 There exists a substantial literature on comparative constructions and their relation with gradable adjectives. Klein (1980, 1982) are classic articles many of whose principles still stand today, and this work has recently been used as the basis of a computer program that is (at least in principle) able to compute the meaning of such sentences as 'Find me a cheap holiday in Slovenia', 'How much more expensive is it to fly with Air France instead of BA?', and 'Is the Hotel Apple further from the beach than the Hotel Orange?' (Pulman 2007). Recent

work in this area includes a number of papers by Chris Kennedy (2001, 2007), who is one of the best-known proponents of the view that gradable adjectives should be modelled using degrees. See also Winter (2005).

10 Phenomena such as hedging, metaphor, and irony are studied in an area known as *linguistic pragmatics*. A well-known textbook on pragmatics is Levinson (1983). An example of linguists' attempts to understand metaphor is Wilson and Carston (2006). The authors discuss the sentence 'Sally is a block of ice', as a description of Sally's character. The idea is to check the dictionary definition of the word 'ice' to look for those properties of ice that can be applied to a person (e.g. unpleasant to interact with, unresponsive to people), and to infer that these properties must be intended. Theories of this nature struggle to explain why one particular set of properties is selected over another. Why, for example, do we not understand the sentence as saying that Sally would wither away when left outside the fridge? For a computational account of some related phenomena, see Markert and Hahn (2002).

11 Frequently, scientists' uncertainty derives from the pitfalls of inferential statistics. See our discussion of statistical significance in Chapter 5.

12 A well-documented overview by the BBC of the different versions of the notorious 45-minutes claim concerning Iraq's weapons of mass destruction can be found on <http://news.bbc.co.uk/2/hi/uk_news/politics/3466005.stm>, in an article called 'Timeline: The 45-minute claim' (as read in October 2008). I have quoted from 'The decision to go to war in Iraq', Ninth Report of Session 2002–3, 3 July 2003, House of Commons Foreign Affairs Committee. The 'early document' was a draft assessment of 5 Sept. 2002 by the Joint Intelligence Committee.

Chapter 7

1 In Conan Doyle's famous *A Study in Scarlet*, the murderer is identified through a chain of inference steps, all of which involve reasoning with vague premisses and/or conclusions: the murderer must be 'in the prime of his life' because he jumped over a 4.5-foot puddle of water; his fingernails are 'remarkably long' because the plaster on the wall on which he wrote something (in blood!) was 'slightly scratched'; the brand of his cigar is Trichinopoly because its ash is 'dark in colour and flaky'. Reasoning along similar lines, his face was inferred

to be 'florid' and his feet 'small for his height'. Having inferred so much about him, Sherlock Holmes has little difficulty finding the perpetrator.

2 We saw that the two laws of comparatives cited in this chapter do not capture what it means for a to be *significantly* more X than b, so the question is how such vague comparatives can be captured. This puzzle was solved in Luce (1956). (See also van Rooij 2009, with roots in van Benthem 1982.) We use a bit of formalism to state Luce's laws here, writing aRb to say that a is significantly more X than b. The first condition is known as *irreflexivity*, the second as the *interval condition*, and the third as *semi-transitivity*. The conditions are meant as sweeping statements; the first, for example, states that for all objects x, xRx is false (i.e. there cannot exist an object that is significantly more X than itself). Relations that fulfil all of (1)–(3) are known as *semi-orders*.

(1) Not (xRx)
(2) If xRy and vRw then xRw or vRy
(3) If xRy and yRz then xRv or vRz

When conditions (1)–(3) are combined, it follows that the relation R must also be transitive (i.e. if xRy and yRz then xRz) and asymmetric (if xRy then not yRx). Two objects a and b are defined to be indistinguishable (as far as the dimension X is concerned) if neither aRb nor bRa holds. Crucially, the conditions (1)–(3) do *not* make indistinguishability transitive, which was Luce's main achievement.

3 For transitivity and related concepts, see any textbook on discrete mathematics (e.g. chapter on 'Relations' in Rosen 2007). Note that in saying that indistinguishability is not transitive, we stop short of saying that it is *intransitive*, which would be the stronger claim that $x \sim y$ and $y \sim z$ forbids $x \sim z$. Notions such as pseudo-connectedness are discussed in Adams (1993); see also Roberts (1979) and van Benthem's contribution to Rubinstein (2000).

4 Arguments against indistinguishability being transitive are at least as old as Luce (1956) but have been questioned in Fara (2001). My own argument which relates indistinguishability to a measurement device does not hinge on any particular properties of the weighing balance that was used. It will go through for any measurement tool, as long as it consistently 'sees' all differences that exceed a certain value, while consistently *not* seeing any differences below some lower value.

5 See the influential Fara (2001).

6 One elegant proof that anything will follow from a contradiction goes as follows: suppose *p* is proven and so is ¬*p*. Given the latter, you may choose any arbitrary proposition *q* and it will be true that (¬*p* or *q*) (because ¬*p* has been proven). But *p* has been proven, therefore (¬*p* or *q*) can be true only in virtue of its right half, which is *q*. In this way we have proven an *arbitrary* proposition *q*, based on nothing but the assumption that both *p* and ¬*p* have been proven. This illustrates why most formal systems collapse as soon as contradictory statements are proven in it.

7 My focus on the split version of the paradox, which involves a long sequence of conditionals, does not involve an important simplification, because any response to one version of the paradox can easily be made to cover the other version as well. Perhaps the only point where the difference between the two versions matters is in connection with *degree theories*. In fuzzy logic, for example, the treatment of universal quantification makes it necessary to use a mathematical *continuum* of truth values rather than, for example, the set of all *fractions* between 0 and 1 which is densely but not continuously ordered. Since other factors make the use of a continuum attractive for fuzzy logic in any case, this is a technicality that will not occupy us.

8 My discussion of the history of sorites owes much to Moline (1969), but the suggestion that Eubulides might have reused the term 'sorites' because of a resemblance with 'Aristotelian Sorites' arguments is my own.

9 The notion of an Aristotelian sorites is discussed in Phillips (1957), which starts with the (now curiously outdated) statement 'Every logician knows the so-called Aristotelian sorites.' There appear to be two possibilities. One is that Aristotle, or others in his time, used the term 'sorites' to designate the (unproblematic) Aristotelian sorites arguments, in which case it seems plausible that Eubulides generalized the term to paradoxical cases involving vagueness. The other, more generally accepted, possibility is that the term 'Aristotelian sorites' came into existence only at some later time, because of the structural similarity between certain syllogisms discussed by Aristotle and Eubulides' paradoxical argument.

10 The comparison between the sorites paradox and the accumulation of error when copying analogue audiotape comes from Pinker (1997: 129).

11 The epistemic perspective on the sorites paradox was defended in Williamson (1994), and in Sorensen (2001); see also Bonini et al. (1999) for psychological experiments and, especially, Tuck (2008) for an application to the 'free riding' problem beloved of political scientists. A discussion of the notion of a paradox is Sorensen (2005). The book is a treasure trove of anecdotes, such as the one in which St Augustine asked what God was doing before He made the world, and was told: 'Preparing hell for people who ask questions like that.'

12 One of the earliest examples of non-classical symbolic logic is the three-valued logic of Łukasiewicz (1920). For a general overview of non-classical logics, see Gabbay and Guenthner (1984). Logic has even produced precisely formalized theories stipulating how logical systems can legitimately differ (e.g. Troelstra 1991). As for the non-classical notion of non-monotonicity, this concept lies right at the heart of the programming language PROLOG. Non-monotonic default inheritance is part of some of the most widely used programming languages, including JAVA. A good example of progress regarding the logical theory of common sense argumentation is Dung (1995), whose influential work does away with the notions of truth and falsity altogether. Perhaps the most important area of non-classical logic at the moment, with applications ranging from computing science to linguistics, is 'dynamic' logic (Harel et al. 2000; also van Benthem 1991, van Ditmarsch 2007).

13 For empirical studies into expressions such as 'evening', see Reiter et al. (2005), which also details a computer program which reduces these problems by using words such as 'evening' with perfect consistency. See also Chapter 11, where computer programs of this kind are discussed.

14 Given Ann's instruction to look for her topology book among the blue books in her collection of 1,000, Parikh derives Bob's expectation thus: perhaps Bob is *lucky*, because the topology book is among the books both she and Bob call blue. The chances of this happening are 0.9 (225/250), in which case Bob has to check about only 150 books on average (300/2). Or perhaps Bob is *unlucky*, because the topology book is one of those Ann calls blue but he does not. If this unlucky scenario (whose probability is 0.1) materializes, then Bob first checks *all* the 300 books he calls blue; after that, lamenting Ann's poor judgement no doubt, he has to check about 350 books on average (half the books in all the

700 he doesn't call blue). Thus, Bob should reasonably expect to have to check about 0.9 times 150, plus 0.1 times (300 + 350), that is 135 + 65, which is 200 books in total, far less than the 500 Bob should expect to check without the benefit of the word 'blue'.

15 Much of the discussion of human colour perception at the start of this chapter is based on Hilbert (1987). For key issues such as colour indistinguishability, Hilbert's book relies on works such as Judd and Wyszecki (1975). For a venerable dictionary of colours and colour terms, see Maerz and Paul (1950).

16 The problems of data sparseness and inconsistency, which form the substance of this dialogue, apply to concepts like that of a 'mountain' with particular force. Mountains often lack commonly agreed boundaries. The rhubarb argument can be repeated in this domain, by naming a previously nameless hill. Epistemicism, as I understand it, would predict that even this newly named hill must have exact boundaries, as soon as the name is coined. I am not sure what epistemicism predicts if and when the landscape changes as a result of vulcanism. Do the outlines of the mountain change immediately, or only as speakers adapt their usage to the changing landscape?

17 Different kinds of continuity are briefly discussed in Körner (1967). Mathematical continuity is not an easy concept to grasp. Perhaps the most straightforward definition is based on Dedekind cuts. Consider a structure involving a set of numbers X ordered by a relation $<$ (i.e. smaller than) in the usual way, and where the ordering is dense. A *Dedekind cut* on X is a way of splitting X into two subsets, where all elements of one subset are smaller than all the elements of the other: for ease of reference, call the set of smaller elements X_{small} and the set of larger elements X_{large}.

Before we define the notion of continuity, consider what happens if a Dedekind cut splits X into a set X_{small} that has no largest element and a set X_{large} that has no smallest element. An example is the situation where X is the set of all fractions, which can be split between the fractions smaller than π and the fractions greater than π. The number π itself is not a fraction, so by missing out on π this split does not leave out any element of X. In this case, the cut can be seen as identifying a kind of *gap* between X_{small} and X_{large}. The gap, in this case, is where π is. Intuitively speaking, the gap shows that X is not as tightly packed as it might have been.

By contrast, consider the situation where X is the set of all real numbers, which includes π. In this case π does *not* define a gap. This is because π itself is no member of either X_{small} or X_{large}, therefore these two sets do not constitute a cut. The definition of continuity now goes as follows. *A structure involving the set X is continuous if it is impossible to define a Dedekind cut on X unless either X_{small} has a largest element or X_{large} has a smallest element.* No gaps can arise, in other words. It turns out that the set of all real numbers is continuous in this sense (see e.g. van Dalen et al. 1978 for a proof). One can split the set into, for example, the real numbers smaller than π and the ones greater than *or equal* to π, but this gives the latter of the two subsets a smallest element, namely π.

Chapter 8

1 Authors notoriously differ in their responses to the sorites paradox but there seems to be broad agreement over what kinds of response are possible in principle (e.g. Hyde 2002, Fara 2000, Williamson 1994). Context-based approaches are often underrepresented, but see van Rooij (2009) for a survey that gives pride of place to them. Many of the crucial older papers on sorites have been brought together in Keefe and Smith (1997).

2 One situation where (1) almost all A are B, (2) almost all B are C, yet (3) *not* almost all A are C is found in the diagram. The situation suggested here is one where A contains 10 elements, 9 of which are elements of B; B contains 91 elements, 82 of which are also elements of C; A and C have no elements at all in common.

3 In keeping with the theme of this book, the borderline between classical logic and other logical systems is not always drawn in exactly the same way by different authors. Fine, for example, regards his own version of partial logic as classical because it uses the same notion of logical consequence as classical logic (e.g. the law of excluded middle holds), even though it uses more than two truth values. The other logical systems discussed in this chapter would probably be regarded as non-classical by every logician.

4 A mathematical theory related to partial logic, with a number of applications in computer science, is Pawlak's theory of rough sets (Pawlak 1991). The core idea is that of approximating a target set, which is somehow difficult to characterize exactly, through a *lower* approximation (i.e. the largest characterizable subset of the target) and an *upper* approximation (i.e. the smallest characterizable

superset of the target). In our example, the set of people whose height is below 150 cm is a lower approximation, and the set of people whose height is below 175 cm an upper approximation.

5 Kit Fine did recognize the limitations of a partial logic account that is able to distinguish between only three kinds of objects in regard to a property such as Short, namely objects x that definitely are short (abbreviated $DShort(x)$), objects that definitely are not short ($D\neg Short(x)$), and objects x which are neither (($\neg DShort(x)$) $\wedge \neg (D\neg Short(x))$). For this reason, his paper explores how partial logic can be extended to deal with *higher-order* vagueness (as when it is unclear whether something is a borderline case or not; see Chapter 1). Although this adds subtlety to the distinctions that can be expressed, the resulting account is not without technical difficulties, and its ability to illuminate vague language appears to be limited. For other discussions of higher-order vagueness, see Wright (1976) and other works by the same author; also Égré and Bonnay (2009) for a perspective using signal detection.

6 The history of the context-based approach is difficult to unravel, because old solutions have sometimes taken a long time to be rediscovered. The proposal by Veltman and Muskens, for example, appears never to have been published except as lecture notes (but see van Deemter 1996 for a summary). Later proposals contained similar ideas, such as the idea that whether or not two stimuli can be distinguished may depend on previous stimuli (e.g. Fara 2001). Even the proposal of Kamp (1981) has taken a long time to become known outside a small circle of logically oriented semanticists. Other context-based approaches to vagueness include Raffman (1996), Soames (1999), Shapiro (2006), and van Rooij (2009). For general discussions of context-dependency and the dynamic turn in linguistics, see van Eijck and Kamp (1997). See van Deemter (1996) for connections between context-dependency and vagueness.

7 Later developments of Hans Kamp's work include an approach to the meaning of sentences and texts known as discourse representation theory (also DRT for short), which is now one of the most detailed versions of the 'dynamic' perspective on meaning, to which the text alludes. An accessible introduction to this wide-ranging theory is contained in Kamp and Reyle (1993).

8 For details of my earlier ambiguity-based approach to sorites, see van Deemter (1991, 1996); also van Rooij (2009). A new approach that likewise

emphasizes the ambiguity of the premises of the argument is sketched in the section on introspective agents, later in this chapter.

9 The paradox of perfect perception is a variant of an argument put forward in Dummett (1975). (See also Luce 1956, Lehrer and Wagner 1985). Dummett's argument involves the hands of a clock and starts from the premiss that space is mathematically continuous. From this premiss he proves that even the crudest observer should be able to distinguish between any two positions of the minute hand, by comparing these positions with a carefully chosen third position. Rather than questioning the crisp notion of indistinguishability (as we do in this section), Dummett uses the argument to conclude that vague predicates give rise to logical inconsistency.

10 It might be thought that, in the approach based on introspective agents, the paradox of perfect perception can be resurrected using the margin of error: Suppose Robo sees two individuals: John's height is 155 cm, while Jon is infinitesimally taller. Given that John is just one MoE below Robo's threshold for shortness, which was 160 cm, one might think that Robo's powers of introspection force it to infer a difference in height between John and Jon. Note, however, that Robo has no access to these individuals' actual heights: it can only measure them and, for all we know, Robo's measurement may (incorrectly) put John above Jon.

Chapter 9

1 Łukasiewicz proposed a three-valued logic (Łukasiewicz 1920) which was later generalized to larger sets of truth values. Max Black's main contributions to many-valued logic include Black (1937, 1949). Lotfi Zadeh's notable publications on the subject include Zadeh (1965) followed by a series of three papers in the journal *Information Sciences* (Zadeh 1975). The mathematical foundations of fuzzy logic have been developed substantially in recent years, including axiomatizations of different versions of the logic (e.g. Hájek 1998, Bergmann 2008), and these have greatly added to mathematicians' respect for this non-classical logic. Until recently it was surprisingly difficult to find good discussions of the sorites paradox in the fuzzy logic literature. The practical handbook Cox (1999) contains a brief discussion on the sorites paradox (called the heap metaphor) on pp. 199–201, focusing on different definitions of logical conjunction.

The paradox, however, plays an important role throughout the more theoretical monograph Bergmann (2008).

2 Definitions using fuzzy logic of words such as 'very' and 'slightly' are often offered without empirical justification (e.g. Negnevitsky 2002, the source of the equations in the text). An exception, where tentative justifications for definitions are discussed, is Zimmermann (1985: Chapter 14).

3 For an informal discussion of these and related statistical blunders, see Goldacre (2009: 271ff.). Goldacre also rightly points out that statistically unlikely events take place all the time, but that's another matter.

Chapter 10

1 One of the best textbooks on artificial intelligence is Russell and Norvig (2003), which contains a detailed survey of all the main concepts and techniques of the field.

2 The example involving the English word 'open' is taken from a standard textbook, Allen (1987).

3 An excellent and very readable biography of Alan Turing is Hodges (1983). In addition to Turing's crucial work on mathematics, code-breaking and artificial intelligence, the book also tells the sad story of a man driven to his death by people who did not approve of his sexual orientation, and 'his lack of reverence for everything except the truth' (ibid. 522).

4 My brief discussion of qualitative reasoning owes much to the chapter on the subject in Bratko (2001). The same chapter is a good source for further references in this area, also somewhat unflatteringly known as *naive physics*. A book focusing entirely on qualitative reasoning is Kuipers (1994).

5 My explanation of Mamdani-style fuzzy inference is largely based on an example presented in Negnevitsky (2002: Chapter 4).

Chapter 11

1 Quoted from Geraats (2007).

2 The standard reference on applied natural language generation (NLG) is Reiter and Dale (2000). The book outlines what is still a widely accepted

division of labour between different generation tasks. NLG systems have also started to reach out to other areas of artificial intelligence, for example by modelling the words, speech, and gestures of a software agent (see van Deemter et al. 2008 for a survey).

3 FoG stands for *forecast generator*. A description of the FoG weather forecasting program is Goldberg et al. (1994). The work on weather reports for road ice is described in Turner et al. (2008, 2009). For modern general-purpose weather forecasting systems that involve natural language, see websites such as the American <http://www.weather.gov/organization.php>, accessed 31 July 2009. Also Reiter et al. (2005).

4 Relevant literature on the (non-)intelligibility of numerical information includes Berry et al. (2002) on systematic misinterpretation of EU-normed expressions of probability, Toogood (1980) on differences in people's interpretation of words such as 'usually', and various publications by Gerd Gigerenzer and colleagues on the explanation of relative risk (e.g. Gigerenzer 2002). For a natural language generation perspective, see Williams and Power (2009).

5 The generation of referring expressions is now one of the best understood areas of NLG. An early seminal paper in this area is Dale and Reiter (1995). The role of salience in the generation of referring expressions is discussed in Krahmer and Theune (2002). The earliest discussion of *vague descriptions* appears to be Pinkal (1979). 'Vague description' is, strictly speaking, a misnomer: in the same way that a 'disabled toilet' is not itself disabled, a vague description, when used appropriately, is not itself vague, even though one of the words in it is vague (i.e. it can give rise to borderline cases).

6 The approach outlined here stems from van Deemter (2006). A computer program called VAGUE, which works along the lines proposed there, was implemented by Richard Power (see <http://www.csd.abdn.ac.uk/~kvdeemte/vague.html>, accessed 31 July 2009). The observation that a reference to an animal as 'the large chihuahua' is appropriate only if the 'gap' between the size of that chihuahua and its nearest competitor is large enough (and at least distinguishable!) is elaborated in the paper, where it is called the *small gaps constraint*; for a related set of observations in connection with the sorites paradox, see van Rooij (2009). For experimental studies concerning the choice between the different forms of the adjective (e.g. 'the large', 'the larger',

'the largest'), see van Deemter (2004). For further developments, see Fernández (2009). For an alternative approach, based on the probability that the reader of a description finds the intended referent, see Horacek (2005). The connection between reference, salience, and pointing is discussed in Kranstedt et al. (2006) and in Piwek (2009).

7 The distinction between local and global vagueness is loosely inspired by the well-known distinction between local and global ambiguity. The word 'letter', for example, is ambiguous between an individual character and an epistle, but in the expression 'love letter', the first word *disambiguates* the second, so it can no longer refer to an individual character. In this two-word expression, 'letter' is therefore locally but not globally ambiguous. The distinction between local and global vagueness is arguably implicit in Kyburg and Morreau (2000).

8 General introductions to game theory include Rasmussen (2001) and Myerson (1991). An intriguing exploration of the connections between game theory and language is Rubinstein (2000), which includes van Benthem (2000) and Lipman (2000). Lipman's chapter, which targets vagueness specifically, is being elaborated in the working paper Lipman (2006); see also van Deemter (2009), which takes a natural language generation perspective on the issue. The work of Rohit Parikh, discussed in Chapter 7, uses a utility-based perspective compatible with game theory (Parikh 1994). Krifka's analysis of rounding in Chapter 7 uses a similar outlook. Connections between game theory, linguistics, and logic are explored in van Benthem (2008). The notion of the *cost* (as separate from the benefits) of an action has come to the fore in publications such as Parikh (2000), van Rooij (2003), and Jäger (2003).

9 The literature does contain one or two examples of formal games that aim to model vagueness. An example is de Jaegher (2003), which involves a more complex version of the game of the two generals. De Jaegher's game lets one general tell the other about the preparedness of the enemy. The utility of vagueness hinges on a subtle asymmetry between the generals, only one of whom will suffer if the enemy turns out to be prepared. Intriguing though it is, I find it difficult to see how de Jaegher's game, which hinges on conflicting interests in quite a subtle way, is relevant to everyday communication or NLG.

10 The work of H. L. A. Hart has caused Waismann's idea of open-texture (e.g. Waismann 1968) to become commonplace among students of law.

Students of language and logic were recently reintroduced to Waismann's work through Shapiro (2006). The example of the Indecent Displays (Control) Act is taken from McLeod (2007). Vagueness in law is discussed extensively in Hart (1994).

11 There exists a substantial psychological literature on the way in which children learn to use and understand vague words. Peccei (1994), for example, points out that vague adjectives such as 'big' are invariably among the first dozens of words that children learn to use. Ebeling and Gelman (1994) describe how typical children of 24 months old are able to understand a variety of fairly complicated vague expressions.

12 Baddeley (2007) discusses a number of mechanisms that might play a role in memories becoming increasingly vague over time, including *semantic coding* (whereby only the meaning of an expression is retained, while the expression itself is forgotten) and *chunking* (whereby a handful of items are clustered together as one unit). There exists a rich literature on memory and forgetting, but I do not know of any psychological research directly addressing the fading (in the sense of becoming vague) of memories over time. The role of memory is discussed in a game theory context by Rubinstein (1998).

13 One possible objection to this seventh response to Lipman is that the flexibility advocated here might be achievable without understanding 'tall' as a vague concept. It would suffice, for example, if the Emperor understood 'tall' as a crisp concept while at the same time being able to reason about other possible definitions of the same concept. Although this situation would be theoretically possible, it would make tallness crisp in name only. The seventh answer to Lipman is worked out in more detail in van Deemter (forthcoming).

14 Experiments such as those in Keysar et al. (2003) and Horton and Keysar (1996) have given rise to a discussion among psycholinguists concerning the extent to which people design their utterances to take their audience into account. This discussion is known as the *egocentricity debate*. For recent confirmation of the idea that speakers' utterances can systematically fall short of what hearers require, see Belz and Gatt (2008). Another example of the combination of hearer-oriented and speaker-oriented experimentation is Paraboni et al. (2006), which focuses on a situation in which an object needs to be identified by means of a referring expression. When the object is a picture in a book, for example,

it might be referred to as 'picture 4' or more elaborately as 'picture 4 in section 2'. In view of the fact that speakers can vary the explicitness of their utterance substantially, two issues are explored: (1) which referring expressions experimental subjects (as authors) prefer to use, and (2) which referring expressions make it easiest for subjects (as readers) to *find* the referent.

Chapter 12

1 Degree theories become particularly tempting once a game-theoretical perspective is adopted, because this perspective focuses on the *utility* of an utterance, and utility is itself a gradable concept. One way to see this is to consider versions of sorites that focus on degree of utility. '*The utility of calling people of 150 cm tall is low. If the utility of calling people of size x tall is low, then the utility of calling people whose height is slightly greater than x tall is also low. Etc.*'

RECOMMENDED READING

The following is a selective, and unashamedly personal, guide to the wider literature associated with this book, divided into a few broad categories.

1. Philosophical issues surrounding vagueness

An obvious area for further reading is the extensive literature on the sorites paradox and related puzzles. If you read just one research paper on vagueness, I recommend Dummett (1975): a model of clarity, this paper makes us realize how unclear and convoluted our thinking about measurement and perception is. For a linguistic perspective on the meaning of vague words, I recommend Kamp (1975). An excellent general resource on philosophical questions surrounding vagueness is Keefe and Smith (1997), which contains much of the older literature (including Dummett's paper), sometimes in slightly updated form. A good and comparatively accessible monograph is Keefe (2000); a classic is Williamson (1994). For a recent survey, which fills some notable gaps in earlier overviews (e.g. by devoting attention to context-based approaches), see van Rooij (2009).

2. Language and maths

Quite apart from vagueness, there exists a substantial literature on mathematical approaches to language, many of which have symbolic logic at their heart. If it's classical logic itself you are interested in, Hunter (1996) is a good start; a solid treatise is Enderton (2001); a computational approach is Fitting (1996). For extensions and variants of classical logic, there is the Handbook series Gabbay and Guenthner (1984). If you're interested in using logic to analyse language, then some excellent and readable introductions will come to your aid, including Partee et al. (1990) and Gamut (1991), neither of which presupposes any mathematical or linguistic background. For a thorough overview of research on logic and language, see van Benthem and ter Meulen (1997). An accessible and wide-ranging textbook on discrete mathematics (including introductory material on symbolic logic, set theory, probability, and computational complexity), often used by students of computing science, is Rosen (2007). For more detailed guidance on set theory, including concepts such as mathematical continuity, I recommend van Dalen et al. (1978); similarly useful is Landman (1991), a textbook that covers many concepts in logic, set theory, and linear algebra that are relevant for vagueness. A venerable mathematical treatise on the theory of measurement is Roberts (1979). For an excellent and accessible introduction to the formal logic of probability (a topic relevant to our Chapter 9), I recommend Adams (1998).

Connections between language and game theory have entered the mainstream of research only in the last decade. An accessible introduction to game theory is Binmore (1991); classic textbooks are Osborne and Rubinstein (1994) and Rasmussen (2001). Links between game theory and language are explored informally in Rubinstein (2000). (At the time of writing, many of Ariel Rubinstein's books are available online, and free of charge, from his home page.) A thorough treatise on many-valued logic, including fuzzy logic, is Bergmann (2008); Zimmermann (1985) is for the more practically inclined, such as engineers interested in fuzzy control and fuzzy set theory. A beautifully informal book on the mathematics behind computing science is Harel (1989).

3. Artificial intelligence and computational linguistics

For a humourous introduction to artificial intelligence, take a look at the cartoon-based booklet Brighton and Selina (2003). A more serious yet easy-going introduction is Cawsey (1997). The most extensive general textbook on artificial intelligence is the excellent Russell and Norvig (2003). A voluminous handbook, written from a cognitive science perspective, is Posner (1989). Finally, although the avowed topic of Hofstadter (1980) was Kurt Gödel's incompleteness theorem and its implications for artificial intelligence, perhaps the main virtue of Hofstadter's Pulitzer Prize-winning bestseller was that it did such a magnificent job of informally explaining the basic concepts of logic and computing, many of which are relevant for the present book as well.

A good overview of computational linguistics, in the form of a handbook, is Mitkov (2003). An excellent though dated textbook is Allen (1987). A widely used textbook representing the current statistical trend in this area of research is Manning and Schütze (1999). An example of a more symbolic approach, detailing the construction of the type of question-answering system discussed in Chapter 6, is Alshawi (1992). The standard reference on applied natural language generation (see our own Chapter 11) is Reiter and Dale (2000), accessible to anyone with a good high-school background. Bateman and Zock (2009) is an up-to-date overview of implemented natural language generation systems.

4. Related themes in popular science

I have had occasion to mention quite a few works that were written for a wide audience and which relate to the subject matter of this book indirectly. This varied collection of books includes Blastland and Dilnot (2008) on abuses of numbers in political discourse; Chang (2007) on the history of temperature measurement; Dawkins (2004) on biology and evolution (see our Chapter 2 on the notion of a species); Gigerenzer (2002) on the psychology of numbers and statistics; Goldacre (2009) on abuses of statistics and on the maltreatment of science in the mass media; Paulos (1995) on various mathematical issues relating to the daily news; Penn and Zalesne (2007) on social trends that go unnoticed because they are slow; Pinker (1997) on human psychology, with plenty of attention to language processing; Weinstein and Adam (2008) on calculating with approximations ('guesstimation'); and Whitelaw (2007), which offers a short history of measurement through the ages. All these books are highly recommended.

REFERENCES

Adams, Ernest, 1993. 'Formalizing the logic of positive, comparative, and superlative'. *Notre Dame Journal of Formal Logic* 34/1: 90–9.

——1998. *A Primer of Probability Logic.* Stanford, Calif.: CSLI.

Allen, James, 1987. *Natural Language Understanding.* Redwood City, Calif.: Benjamin Cummings. 2nd edn. 1994.

Alshawi, Hiyan (ed.), 1992. *The Core Language Engine.* Cambridge, Mass.: MIT.

Aragonès, Enriqueta, and Zvika Neeman, 2000. 'Strategic ambiguity in electoral competition', *Journal of Theoretical Politics* 12: 183–204.

Baddeley, Alan, 2007. *Working Memory, Thought, and Action.* Oxford: Oxford University Press.

Bar-Hillel, Yehoshua, 1966. 'The present status of automatic translation of languages'. In *Advances in Computers* 1: 91–163.

Barker, Chris, 2002. 'The dynamics of vagueness', *Linguistics and Philosophy* 25/1: 1–36.

Bartsch, Renate, and Theo Vennemann, 1983. *Grundzüge der Sprachtheorie: eine Linguistische Einfuerung* (Principles of Language Theory: A Linguistic Introduction). Tübingen: Max Niemeyer.

Bateman, John, and Michael Zock, 2009. 'John Bateman and Michael Zock's list of Natural Language Generation Systems'. Downloadable from <http://www.fb10.uni-bremen.de/anglistik/langpro/NLG-table/NLG-table-root.htm>, accessed 23 April 2009.

Belz Anja, and Albert Gatt, 2008. 'Intrinsic vs. extrinsic evaluation measures for referring expressions generation'. *Proceedings of the 46th Annual Meeting of the Association for Computational Linguistics* (ACL-2008), Columbia, Ohio, 197–200.

Bergmann, Merrie, 2008. *An Introduction to Many-Valued and Fuzzy Logic: Semantics, Algebras, and Derivation Systems.* Cambridge: Cambridge University Press.

REFERENCES

Berry, Dianne, Peter Knapp, and Theo Raynor, 2002. 'Is 15 percent very common? Informing people about the risks of medication side effects', *International Journal of Pharmacy Practice* 10: 145–51.

Binmore, Kenneth, 1991. *Fun and Games: A Text on Game Theory*. Lexington, Mass.: D. C. Heath and Co.

Black, Max, 1937. 'Vagueness: An exercise in logical analysis'. In Black (1949), 25–58.

——1949. *Language and Philosophy*. Ithaca, NY: Cornell University Press.

Blastland, Michael, and Andrew Dilnot, 2008. *The Tiger that Isn't: Seeing Through a World of Numbers*. 2nd, expanded edn. London: Profile Books.

Bonini, Nicolao, Daniel Osherson, Riccardo Viale, and Timothy Williamson, 1999. 'On the psychology of vague predicates', *Mind and Language* 14/4: 377–93.

Bratko, Ivan, 2001. *Prolog: Programming for Artificial Intelligence*. 3rd edn. Harlow: Addison-Wesley.

Brighton, Henry, and Howard Selina, 2003. *Introducing Artificial Intelligence*. Cambridge: Icon Books.

Bronnenberg, Wim, Harry Bunt, Jan Landsbergen, Remko Scha, Wijnand Schoenmakers, and Eric van Utteren, 1980. 'The Question-Answering System PHLIQA1'. In L. Bolc (ed.), *Natural Language Question Answering Systems*. Munich: Hanser; London: Macmillan, 217–305.

Cawsey, Alison, 1997. *The Essence of Artificial Intelligence*. Englewood Cliffs, NJ: Prentice Hall.

Chang, Hasok, 2007. *Inventing Temperature: Measurement and Scientific Progress*. Oxford: Oxford University Press.

Chomsky, Noam, 1965. *Aspects of a Theory of Syntax*. Cambridge, Mass.: MIT.

Cole, Tim J., et al., 2000. 'Establishing a standard definition for child overweight and obesity worldwide: International survey'. *British Medical Journal* 320: 1240–3.

Conan Doyle, Arthur, 1888. *A Study in Scarlet*. London: Ward Lock and Co. First pub. in *Beeton's Christmas Annual*, 1887. Repr. as a Penguin Classic, 2001.

Cox, Earl, 1999. *The Fuzzy Systems Handbook: A Practitioner's Guide to Building, Using, and Maintaining Fuzzy Systems*. San Diego: Academic Press.

Crawford, V., and J. Sobel, 1982. 'Strategic information transmission', *Econometrica* 50: 1431–51.

Dale, R., and E. Reiter, 1995. 'Computational interpretations of the Gricean maxims in the generation of referring expressions', *Cognitive Science* 19: 233–63.

REFERENCES

Dawkins, Richard, 2004. *The Ancestor's Tale: A Pilgrimage to the Dawn of Life.* Weidenfeld & Nicolson. Paperback edn., London: Phoenix Orion, 2005.

de Jaegher, Kris, 2003. 'A game-theoretical rationale for vagueness', *Linguistics and Philosophy* 26: 637–59.

De Luc, Jean-André, 1772. *Recherches sur les modification de l'atmosphère* (Research on Modifications of the Atmosphere). 2 vols. Geneva: n.p. Also published in Paris by La Veuve Duchesne, in 4 vols., in 1784 and 1778.

DeVault, David, and Matthew Stone, 2004. 'Interpreting vague utterances in context', *Proceedings of the 20th International Conference on Computational Linguistics* (COLING-2004), Geneva, Switzerland (Art. 1247).

De Vet, John H. M., 1988. 'A practical algorithm for evaluating database queries', *Software: Practice and Experience* 19/5: 491–504.

Dickens, William T., and James R. Flynn, 2006. 'Black Americans reduce the racial IQ gap: Evidence from standardization samples', *Psychological Science*, Oct.

Dobzhansky, Theodosius, 1937. *Genetics and the Origin of Species.* New York: Columbia University Press.

Drake, Frank R., 1974. *Set Theory: An Introduction to Large Cardinals.* Studies in Logic and Foundations of Mathematics 76. Amsterdam: North-Holland.

Dummett, Michael, 1975. 'Wang's Paradox', *Synthese* 30: 301–24. Repr. in Keefe and Smith (1997).

Dung, Phan Minh, 1995. 'On the acceptability of arguments and its fundamental role in nonmonotonic reasoning, logic programming, and n-person games', *Artificial Intelligence* 7/2: 321–57.

Ebeling, K. S., and S. A. Gelman, 1994. 'Children's use of context in interpreting "big" and "little"', *Child Development* 65/4: 1178–92.

Edgington, Dorothy, 1992. 'Validity, uncertainty and vagueness', *Analysis* 52: 193–204.

——1996. 'Vagueness by degrees'. In Keefe and Smith (1997).

Égré, Paul, and Denis Bonnay, 2009. 'Inexact knowledge with introspection', *Journal of Philosophical Logic* 38/2: 179–228.

Enderton, Herbert B., 2001. *A Mathematical Introduction to Logic.* 2nd edn. Burlington, Mass.: Harcourt/Academic Press.

Essenin-Volpin, A. S., 1970. 'The ultra-intuitionistic criticism and the anti-traditional program for foundations of mathematics'. In A. Kino, John Myhill, and

E. Vesley (eds.), *Intuitionism and Proof Theory*. Proceedings of the Summer Conference at Buffalo, New York, 1968. Amsterdam: North-Holland, 3–45.

Fara, Delia Graff, 2000. 'Shifting sands: An interest-relative theory of vagueness', *Philosophical Topics* 28/1: 45–81. Originally published under the name 'Delia Graff'.

——2001. 'Phenomenal continua and the sorites', *Mind* 110/440: 905–36. Originally published under the name 'Delia Graff'.

Fernández, Raquel, 2009. 'Salience and feature variability in definite descriptions with positive-form vague adjectives'. In *Proceedings of COGSCI 2009, Workshop 'Production of Referring Expressions: Bridging the Gap between Empirical and Computational Approaches to Reference'*. Amsterdam.

Fine, Kit, 1975. 'Vagueness, truth and logic', *Synthese* 30: 265–300.

Fishburn, Peter C., 1986. 'Axioms of subjective probability', *Statistical Science* 1/3: 335–58.

Fitting, Melvin, 1996. *First-Order Logic and Automated Theorem Proving*. 2nd edn. New York: Springer.

Forbes, Graeme, 2007. 'Identity and the facts of the matter'. Presentation at the Arché conference, University of St. Andrews.

Gabbay, Dov, and Franz Guenthner (eds.), 1984. *Handbook of Philosophical Logic*, vols. i–xii. Dordrecht: Reidel.

Gamut L. T. F., 1991. *Logic, Language and Information*. 2 vols. Chicago: University of Chicago Press.

Geraats, Petra M., 2007. 'The mystique of Central Bank speak', *International Journal of Central Banking* 3/1: 37–80.

Gigerenzer, Gerd, 2002. *Reckoning with Risk*. London: Penguin.

Goldacre, Ben, 2009. *Bad Science*. London: Harper Perennial.

Goldberg, Eli, Norbert Driedger, and Richard Kittredge, 1994. 'Using Natural-Language Processing to Produce Weather Forecasts', *IEEE Expert* 9/2: 45–53.

Goodman, Nelson, 1954. *Fact, Fiction and Forecast*. London: Athlone.

Grabe, William, and Robert B. Kaplan, 1997. 'On the writing of science and the science of writing: Hedging in science text and elsewhere'. In Raija Markkanen and Hartmut Schroeder (eds.), *Hedging and Discourse: Approaches to the Analysis of a Pragmatic Phenomenon in Academic Texts*. Berlin: Walter de Gruyter, 151–67.

Grice, Paul, 1975. 'Logic and conversation'. In P. Cole and J. L. Morgan, *Syntax and Semantics*, iii. *Speech Acts*. New York: Academic Press, 41–58.

REFERENCES

Hájek, Petr, 1998. *Metamathematics of Fuzzy Logic*. Boston: Kluwer.

Halpern, Joseph, 2004. 'Intransitivity and vagueness'. In *Proceedings of the Ninth International Conference on Principles of Knowledge Representation and Reasoning* (KR-2004), 121–9.

Harel, David, 1989. *The Science of Computing: Exploring the Nature and Power of Algorithms*. Reading, Mass.: Addison-Wesley.

——Dexter Kozen, and Jerzy Tiuryn, 2000. *Dynamic Logic*. Cambridge, Mass.: MIT.

Hart, Herbert Lionel Adolphus, 1994. *The Concept of Law*. Oxford: Clarendon.

Hilbert, David R., 1987. *Color and Color Perception: A Study in Anthropocentric Realism*. CSLI Lecture Notes 9. Stanford, Calif.: CSLI Publications.

Hodges, Andrew, 1983. *Alan Turing: The Enigma of Intelligence*. Boston: Unwin Paperbacks.

Hofstadter, Douglas, 1980. *Gödel, Escher, Bach: An Eternal Golden Braid*. New York: Vintage.

Horacek, Helmut, 2005. 'Generating referential descriptions under conditions of uncertainty'. In *Proceedings of the 10th European Workshop on Natural Language Generation* (ENLG-2005), Aberdeen, Scotland, 58–67.

Hornstein, Norbert, 1998. 'Noam Chomsky'. In Edward Craig (ed.), *Routledge Encyclopaedia of Philosophy*. London: Routledge.

Horton, William S., and Boaz Keysar, 1996. 'When do speakers take into account common ground?' *Cognition* 59: 91–117.

Hripcsak, George, Noémie Elhadad, Cynthia Chen, Li Zhou, and Frances P. Morrison, 2009. 'Using empirical semantic correlation to interpret temporal assertions in clinical texts', *Journal of the American Medical Informatics Society* (JAMIA) 16: 220–7.

Hunter, Geoffrey, 1996. *Metalogic: An Introduction to the Metatheory of Standard First Order Logic*. Berkeley and Los Angeles: University of California Press. (Sixth printing with corrections.)

Huygens, Christiaan, 1673. *Horologium oscillatorium sive de motu pendulorum ad horologia aptato* (The Pendulum Clock, or, the Motion of Pendulums Adapted to Clocks). Paris: F. Muguet. Electronic version available from Cornell University Library, Ithaca, NY.

Hyde, Dominique, 2002. 'Sorites paradox', *Stanford Encyclopedia of Philosophy*. Stanford, Calif.: Stanford University Press. Downloadable from <http://plato.stanford.edu/archives/fall2002/entries/sorites-paradox>, accessed 5 Aug. 2009.

323

Jäger, Gerhard, 2003. 'Applications of game theory in linguistics'. *Language and Linguistics Compass* 2/3: 1749–67.

Janssen, Theo M. V., 1997. 'Compositionality'. In van Benthem and ter Meulen (1997).

Jensen, Arthur, 1979. *Bias in Mental Testing*. New York: Free Press.

Johnson, Robert, and Patricia Kuby, 2000. *Elementary Statistics*. 8th edn. Pacific Grove, Calif.: Duxbury.

Jud, Deane B., and Gunter Wyszecki, 1975. *Color in Business, Science and Industry*. New York: John Wiley & Sons.

Kamp, Hans, 1975. 'Two theories of adjectives'. In E. L. Keenan (ed.), *Formal Semantics of Natural Language*. Cambridge: Cambridge University Press, 123–55.

—— 1981. 'The paradox of the heap'. In Uwe Moennich (ed.), *Aspects of Philosophical Logic*. Dordrecht: Reidel, 225–77.

—— and Uwe Reyle, 1993. *From Discourse to Logic: An Introduction to Model-Theoretic Semantics of Natural Language, Formal Logic and Discourse Representation Theory*. Dordrecht: Kluwer.

Keefe, Rosanna, 2000. *Theories of Vagueness*. Cambridge: Cambridge University Press.

—— and Peter Smith (eds.), 1997. *Vagueness: A Reader*. Cambridge, Mass.: MIT.

Kennedy, Chris, 2001. 'Polar opposition and ontology of "degrees"'. *Linguistics and Philosophy* 24: 33–70.

—— 2007. 'Vagueness and grammar: The semantics of relative and absolute gradable adjectives', *Linguistics and Philosophy* 30/1: 1–45.

Keysar, Boaz, Shuhong Lin, and Dale J. Barr, 2003. 'Limits on theory of mind use in adults', *Cognition* 89: 25–41.

Klein, Ewan, 1980. 'A semantics for positive and comparative adjectives', *Linguistics and Philosophy* 4: 1–45.

—— 1982. 'The interpretation of adjectival comparatives', *Journal of Linguistics* 18: 113–36.

Körner, Stephan, 1967. 'Continuity'. In Paul Edwards (ed.), *Encyclopaedia of Philosophy*, vol. ii. New York: Macmillan.

Krahmer, Emiel, and Mariët Theune, 2002. 'Efficient context-sensitive generation of referring expressions'. In K. van Deemter and R. Kibble (eds.),

Information Sharing: Reference and Presupposition in Language Interpretation and Generation. Stanford: CSLI Publications, 223–64.

Kranstedt A., A. Lücking, T. Pfeiffer, H. Rieser, and I. Wachsmuth, 2006. 'Deiktic object reference in task-oriented dialogue'. In G. Rickheit and I. Wachsmuth (eds.), *Situated Communication*. Berlin: Mouton de Gruyter, 155–208.

Kreisel, Georg, 1987. 'Church's thesis and the ideal of informal rigour', *Notre Dame Journal of Formal Logic* 28/4: 499–519.

Krifka, Manfred, 2002. 'Be brief and be vague!' In David Restle and Dietmar Zaefferer (eds.), *Sounds and Systems: Studies in Structure and Change: A Festschrift for Theo Vennemann*. Trends in Linguistics 141. Berlin: Mouton de Gruyter, 439–58.

Kuipers, Benjamin, 1994. *Qualitative Reasoning: Modeling and Simulation with Incomplete Knowledge*. Cambridge, Mass.: MIT.

Kyburg, Alice, and Michael Morreau, 2000. 'Fitting words: Vague language in context', *Linguistics and Philosophy* 23: 577–97.

Labov, William, 1972. *Language in the Inner City: Studies in Black English Vernacular*. Philadelphia: University of Pennsylvania Press.

Lakoff, George, 1970. 'A note on vagueness and ambiguity', *Linguistic Inquiry* 1.

Landman, Fred, 1991. *Structures for Semantics*. Dordrecht: Kluwer Academic Publishers.

Lehmann, Fritz, and Anthony G. Cohn, 1994. 'The EGG/Yolk reliability hierarchy: Semantic data integration using sorts with prototypes'. In *Proceedings of the Third International Conference on Information and Knowledge Management*, Gaithersburg, Md., 272–9.

Lehrer, Keith, and Carl Wagner, 1985. 'Intransitive indifference: The semi-order problem'. *Synthese* 65: 249–56.

Levinson, Stephen C., 1983. *Pragmatics*. Cambridge: Cambridge University Press.

Lewis, David, 1969. *Convention—A Philosophical Study*. Cambridge, Mass.: Harvard University Press.

Lipman, Barton L., 2000. 'Economics and Language'. In Rubinstein (2000: Pt. III, Comments).

—— 2006. 'Why is language vague?' Working paper, December 2006, Department of Economics, Boston University.

Logvinenko, A. D., W. Byth, and E. E. Vityaev, 2003. 'In search of an elusive hard threshold: A test of observer's ability to order sub-threshold stimuli', *Vision Research* 44/3: 287–96.

REFERENCES

Luce, R. D., 1956. 'Semiorders and a theory of utility discrimination', *Econometrica* 24: 178–91.

Łukasiewicz, Jan, 1920. 'On three-valued logic'. English version in L. Borkowski (ed.), *Selected Works by Jan Łukasiewicz*. Amsterdam: North-Holland, 1970.

Maerz, A., and M. Rea Paul, 1950. *A Dictionary of Color*. New York: McGraw-Hill.

Manning, Chris, and Hinrich Schütze, 1999. *Foundations of Statistical Natural Language Processing*. Cambridge, Mass.: MIT.

Marcus, Gary, 2008. *Kluge: The Haphazard Construction of the Human Mind*. London: Faber and Faber.

Markert, Katja, and Udo Hahn, 2002. 'Metonymies in discourse', *Artificial Intelligence* 135/1–2: 145–98.

Mayr, Ernst, 1942. *Systematics and the Origin of Species*. New York: Columbia University Press.

McLeod, Ian, 2007. *Legal Theory*. 4th edn. Basingstoke: Palgrave Macmillan Law Masters.

Mitkov, Ruslan (ed.), 2003. *Oxford Handbook of Computational Linguistics*. Oxford: Oxford University Press.

Moline, Jon, 1969. 'Aristotle, Eubulides and the sorites', *Mind*, NS 78/311: 393–407.

Montague, Richard, 1974. *Formal Philosophy: Selected Papers of Richard Montague*, ed. and intro. Richmond H. Thomason. New Haven: Yale University Press.

Morreau, Michael, 2002. 'What vague objects are like', *Journal of Philosophy* 23/6: 577–9.

Moxey, L. M., and A. J. Sanford, 2000. 'Communicating quantities: A review of psycholinguistic evidence of how expressions determine perspectives', *Applied Psychology* 14/3: 237–55.

Myerson, R. B., 1991. *Game Theory: Analysis of Conflict*. Cambridge, Mass.: Harvard University Press.

Negnevitsky, Michael, 2002. *Artificial Intelligence: A Guide to Intelligent Systems*. Harlow: Addison Wesley.

Neisser, U., 1997. 'Rising scores on intelligence tests', *American Scientist* 85: 440–7.

Osborne, Martin J., and Ariel Rubinstein, 1994. *A Course in Game Theory*. Cambridge Mass.: MIT.

Pallen, Mark, 2009. *The Rough Guide to Evolution*. London: Rough Guides.

Pan, Philip, 2008. *Out of Mao's Shadow: The Struggle for the Soul of a New China*. London: Picador.

REFERENCES

Paraboni, Ivandré, Kees van Deemter, and Judith Masthoff, 2006. 'Generating referring expressions: Making referents easy to identify', *Computational Linguistics* 33/2: 229–54.

Parikh, Rohit, 1994. 'Vagueness and utility: The semantics of common nouns', *Linguistics and Philosophy* 17: 521–35.

——2000. 'Communication, meaning, and interpretation', *Linguistics and Philosophy* 23: 185–212.

Parsons, Terence, 2000. *Indeterminate Identity: Metaphysics and Semantics*. Oxford: Clarendon.

Partee, Barbara H., Alice ter Meulen, and Robert E. Wall, 1990. *Mathematical Methods in Linguistics*. Dordrecht: Kluwer Academic.

Paulos, John Allen, 1995. *A Mathematician Reads the Newspaper*. New York: Random House.

Pawlak, Zdzilaw, 1991. *Rough Sets: Theoretical Aspects of Reasoning about Data*. Dordrecht: Kluwer Academic.

Peccei, Jean Stilwell, 1994. *Child Language*. London: Routledge.

Penn, Mark J., and E. Kinney Zalesne, 2007. *Microtrends*. London: Penguin.

Phillips, H. B., 1957. 'A discovery in traditional logic', *Mind* NS 66/263: 398–400.

Pinkal, Manfred, 1979. 'How to refer with vague descriptions'. In R. Bäuerle, U. Egli, and A. von Stechow (eds.), *Semantics from Different Points of View*. Berlin: Springer, 32–50.

——1995. *Logic and Lexicon*. Oxford: Oxford University Press.

Pinker, Steven, 1997. *How the Mind Works*. London: Penguin.

Piwek, Paul, 2009. 'Salience and pointing in multimodal reference'. In *Proceedings of COGSCI 2009, Workshop 'Production of Referring Expressions: Bridging the Gap between Empirical and Computational Approaches to Reference'*, Amsterdam.

Popper, Karl R., 1959. *The Logic of Scientific Discovery*. New York: Basic Books. First published as *Logik der Forschung*, Vienna, 1934.

Posner, Michael I. (ed.), 1989. *Foundations of Cognitive Science*. Cambridge, Mass.: MIT.

Ptacek, Margaret B., and Shale J. Hankison. 2009. 'The pattern and process of speciation'. In Michael Ruse and Joseph Travis (eds.), *Evolution: The First Four Billion Years*. Cambridge, Mass.: Harvard University Press, 177–207.

Pulman, Stephen, 2007. 'Formal and computational semantics: A case study'. In *Proceedings of Seventh International Workshop on Computational Semantics* (IWCS-7), Tilburg, The Netherlands, 181–96.

REFERENCES

Putnam, Hilary, 1975–83. *Philosophical Papers*, i. *Mind, Language and Reality*; ii. *Meaning and the Moral Sciences*; and iii. *Reason, Truth, and History*. Cambridge: Cambridge University Press.

Quetelet, L. A. J., 1871. *Antropométrie ou mesure des différences facultés de l'homme* (Anthropometry or the Measurement of Different Abilities in People). *Brussels: Musquardt.*

Quirk, Randolph, Sidney Greenbaum, Geoffrey Leech, and Jan Svartvik, 1972. *A Grammar of Contemporary English*. Harlow: Longman.

Raffman, D., 1996. 'Vagueness and context-relativity', *Philosophical Studies* 81: 175–92.

Rasmussen, Eric, 2001. *Games & Information: An Introduction to Game Theory*. 3rd edn. Oxford: Blackwell.

Ray, Debraj, 1998. *Development Economics*. Princeton: Princeton University Press.

Reiter, Ehud, and Robert Dale, 2000. *Building Natural Language Generation Systems*. Cambridge: Cambridge University Press.

Reiter, Ehud, S. Sripada, J. Hunter, J. Yu, and I. Davy, 2005. 'Choosing words in computer-generated weather forecasts', *Artificial Intelligence* 167: 137–69.

Roberts, F., 1979. *Measurement Theory*. Reading, Mass.: Addison-Wesley.

Rosch, Eleanor, 1978. 'Principles of categorization'. In E. Rosch and B. Lloyd (eds.), *Cognition and Categorization*. Hillsdale, NJ: Lawrence Erlbaum, 27–48.

Rosen, Kenneth, 2007. *Discrete Mathematics and its Applications*. 6th edn. Boston: McGraw-Hill.

Rosetta, M. T., 1994. *Compositional Translation*. International Series in Engineering and Computer Science. Dordrecht: Kluwer Academic.

Rubinstein, Ariel, 1998. *Modeling Bounded Rationality*. Cambridge, Mass.: MIT.

—— 2000. *Economics and Language: Five Essays*. Cambridge: Cambridge University Press.

Rucker, Rudy, 1995. *Infinity and the Mind: The Science and Philosophy of the Infinite*. Princeton: Princeton University Press.

Russell, Bertrand, 1923. 'Vagueness', *Australasian Journal of Psychology and Philosophy* 1: 84–92.

—— 1948. *Human Knowledge: Its Scope and Limits*. London: Allen & Unwin.

Russell, Stuart, and Peter Norvig, 2003. *Artificial Intelligence: A Modern Approach*. 2nd edn. Upper Saddle River, NJ: Prentice Hall.

Sainsbury, Mark, 1990. *Concepts without Boundaries*. London: King's College. Repr. in Keefe and Smith (1997).

REFERENCES

Shapiro, Stewart, 2006. *Vagueness in Context*. Oxford: Clarendon.

Shelley, Mary, 1818. *Frankenstein*. London: Lackington, Hughes, Harding, Mavor & Jones.

Skolem, Thoralf, 1920. 'Logico-combinatorial investigations in the satisfiability or provability of mathematical propositions: A simplified proof of a theorem by L. Loewenheim and generalisations of the theorem'. Repr. in Jean van Heijenoort (ed.), *From Frege to Gödel: A Source Book in Mathematical Logic, 1879–1931*. Cambridge, Mass.: Harvard University Press, 1967, 252–63.

Soames, S., 1999. *Understanding Truth*. New York: Oxford University Press.

Sorensen, Roy, 2001. *Vagueness and Contradiction*. Oxford: Oxford University Press.

——2005. *A Brief History of the Paradox: Philosophy and the Labyrinths of the Mind*. Oxford: Oxford University Press.

Stebbins, R. C., 1949. 'Speciation in salamanders of the plethodontic genus *Ensatina*', *University of California Publications in Zoology* 48: 377–526.

Thorniley, Tessa, 2007. 'Will anyone stop the rise of Britain's super-rich?' *Independent on Sunday*, 2 Sept. 2007.

Timbrell, John, 2005. *The Poison Paradox*. Oxford: Oxford University Press.

Toogood, J. H., 1980. 'What do we mean by "usually"?' *Lancet* 1: 1094.

Troelstra, Anne, 1991. *Lectures on Linear Logic*. Stanford, Calif.: CSLI Publications.

Tuck, Richard, 2008. *Free Riding*. Cambridge, Mass.: Harvard University Press.

Turing, Alan, 1950. 'Computing machinery and intelligence', *Mind* 70/236: 433–60.

Turner, R., S. Sripada, E. Reiter, and I. P. Davy. 2008. 'Using spatial reference frames to generate grounded textual summaries of georeferenced data'. In *Proceedings of INLG-2008*. Salt Fork, Ohio, 16–24.

Turner, R., Yaji Sripada, and E. Reiter, 2009. 'Generating approximate geographic descriptions'. In *Proceedings of 12th European Workshop on Natural Language Generation* (ENLG-2009), Athens, Greece, 42–9.

van Benthem, Johan, 1982. 'Later than late: On the logical origin of the temporal order', *Pacific Philosophical Quarterly* 63: 193–203.

——1991. *Language in Action: Categories, Lambdas and Dynamic Logic*. Studies in Logic and the Foundations of Mathematics 130. Amsterdam: North-Holland.

——2000. 'Economics and language'. In Rubinstein (2000: Pt. III, Comments).

——2008. 'Games that makes sense: Logic, language, and multi-agent interaction'. In K. Apt and R. van Rooij (eds.), *Proceedings of KNAW Colloquium on*

329

Games and Interactive Logic. Texts in Logic and Games. Amsterdam: Amsterdam University Press, 197–209.

van Benthem, Johan, and Alice ter Meulen, 1997. *Handbook of Logic and Language*. Cambridge, Mass.: MIT.

van Dalen, Dirk, H. C. Doets, and H. C. M. de Swart, 1978. *Sets, Naive, Axiomatic and Applied*. Oxford: Pergamon Press.

van Deemter, Kees, 1991. 'The role of ambiguity in the sorites fallacy'. In Dekker and Stokhof (eds.), *Proceedings of 9th Amsterdam Colloquium*. Amsterdam: University of Amsterdam, 209–27.

——1996. 'The sorites fallacy and the context-dependence of vague predicates'. In M. Kanazawa, C. Piñon, and H. de Swart (eds.), *Quantifiers, Deduction, and Context*. Stanford, Calif.: CSLI Publications, 59–86.

——1998. 'Ambiguity and idiosyncratic interpretation', *Journal of Semantics* 15/1: 5–36.

——2004. 'Finetuning an NLG system through experiments with human subjects: The case of vague descriptions'. In *Proceedings of the 3rd International Conference on Natural Language Generation* (INLG-04). Brockenhurst, 31–40.

——2006. 'Generating referring expressions that involve gradable properties', *Computational Linguistics* 32/2.

——2009. 'What game theory can do for NLG: The case of vague language'. In *Proceedings of the 12th European Workshop on Natural Language Generation* (ENLG-2009), Athens, Greece.

——(forthcoming). 'Utility and language generation: The case of vagueness', *Journal of Philosophical Logic* 38/6, 607–632.

——and Stanley Peters (eds.), 1996. *Semantic Ambiguity and Underspecification*. Stanford, Calif.: CSLI Publications.

——Brigitte Krenn, Paul Piwek, Marc Schröder, Martin Kleesen, and Stephan Baumann, 2008. 'Fully generated scripted dialogue', *Artificial Intelligence* 172/10: 1219–44.

van der Sluis, Ielka, and Chris Mellish, 2009. 'Towards empirical evaluation of affective tactical NLG'. In *Proceedings of the 12th European Workshop on Natural Language Generation* (ENLG-2009), Athens, Greece, 146–53.

van Ditmarsch, Hans, Wiebe van der Hoek, and B. P. Kooi, 2007. *Dynamic Epistemic Logic*. Synthese Library 337. Berlin: Springer.

van Eijck, Jan, and Hans Kamp, 1997. 'Representing discourse in context'. In van Benthem and ter Meulen (1997), 179–237.

REFERENCES

van Fraassen, Bas, 1966. 'Singular terms, truth value gaps and free logic', *Journal of Philosophy* 63: 481–95.

van Rooij, Robert, 2003. 'Being polite is a handicap: Towards a game theoretical analysis of polite linguistic behavior'. In *Proceedings of Theoretical Aspects of Rationality and Knowledge* (TARK-9), Bloomington, Ind., 45–58.

—— 2009. 'Vagueness and linguistics'. In G. Ronzitti (ed.), to appear in *The Vagueness Handbook*. Dordrecht: Springer.

Veltman, Frank, 2002. 'Het verschil tussen "vaag" en "niet precies"' (*The difference between 'vague' and 'not precise'*). Inaugural lecture, University of Amsterdam. Amsterdam: Vossiuspers.

von Neumann, John, and Oskar Morgenstern, 1944. *Theory of Games and Economic Behavior*. Princeton, NJ: Wiley & Sons.

Waismann, Friedrich, 1968. 'Verifiability'. In Antony Flew (ed.), *Logic and Language*. Oxford: Basil Blackwell, 119–23.

Wake, D. B., 1997. 'Incipient species formation in salamanders of the Ensatina complex', *Proceedings of the National Academy of Sciences of the USA* 94: 7761–67.

Weinstein, Lawrence, and John A. Adam, 2008. *Guesstimation: Solving the World's Problems on the Back of a Cocktail Napkin*. Princeton: Princeton University Press.

Whitelaw, Ian, 2007. *A Measure of All Things: The Story of Measurement through the Ages*. Cincinnati, Ohio: David & Charles.

Williams, Sandra, and Richard Power, 2009. 'Precision and mathematical form in first and subsequent mentions of numerical facts and their relation to document structure'. In *Proceedings of the 12th European Workshop on Natural Language Generation* (ENLG-2009), Athens, Greece, 118–21.

Williamson, Timothy, 1994. *Vagueness*. London: Routledge.

Wilson, Deirdre, and Robyn Carston, 2006. 'Metaphor, relevance and the "emergent property" issue', *Mind and Language* 21/3: 404–33.

Winter, Yoad, 2005. 'Cross-categorial restrictions on measure phrase modifications', *Linguistics and Philosophy* 28: 233–67.

Woods, William, A., 1968. 'Procedural semantics for a question-answering machine'. In *Proceedings of the AFIPS' Fall Joint Computer Conference*, New York, 457–71.

World Health Organization, 2000. *Obesity: Preventing and Managing the Global Epidemic*. Technical Report Series 894. Geneva: World Health Organization.

REFERENCES

Wright, Crispin, 1976. 'Language-mastery and the sorites paradox'. In G. Evans and J. McDowell (eds.), *Truth and Meaning: Essays in Semantics*, Oxford: Clarendon, 223–47. Repr. in Keefe and Smith (1997).

Zadeh, Lotfi, 1965. 'Fuzzy sets', *Information and Control* 8/3: 338–53.

——1975. 'The concept of a linguistic variable and its application to approximate reasoning (parts I, II, and III)', *Information Sciences* 8 and 9.

Zimmermann, Hans J., 1985. *Fuzzy Set Theory — and its Applications*. Boston: Kluwer/Nijhoff.

INDEX

INDEX

Lipman, Barton L. 261–3, 265, 269, 271, 297, 311–2, 325
Löb, Martin 71
Loebner, Hugh 226
logic 13, 25, 28–9, 74, 108, 120, 126–218, 230, 282, 287, 303–4, 306–8, 311–2, 315–17 classical 11, 13, 139, 153, 156–66, 173, 182, 185, 188–9, 190, 193, 198, 200, 210, 215, 229–30, 270, 282–3, 306, 315 conditional 136, 157–8, 160–1, 165–6, 171–2, 175, 200–3, 208–9, 211–3, 217–8, 230, 303 conjunction 157, 161, 166, 172, 194–5, 201–2, 209, 213, 216–7, 308 deviant 139 disjunction 157, 195, 200, 204–5, 231–2 dynamic 304 implication 136, 157 linear 329 logical connective 156, 203, 235 logical consequence 29, 160–1, 252, 306 logical operator 156–7, 216, 218 logical validity 13, 137, 154–5, 159, 160–2, 183, 185, 195, 209 many-valued 187–218, 308, 316 negation 134, 193, 209 non-classical logic 13, 215, 304, 308 non-monotonic 139, 304 partial logic 161–7, 188–9, 271, 282–3, 306–7 predicate logic 159–60, 270 probabilistic 189, 204–14, 283, 287 propositional logic 158–9 symbolic logic 13–4, 155–6, 174, 208, 215, 252, 284, 304, 315 three-valued 304, 308
Logvinenko, A. D. 296, 325
Luce, R. D. 302, 308, 326
Lücking, T. 325
Łukasiewicz, Jan 190, 326
lying 284

Maerz, A. 305, 326
magnitude estimation 198
Mamdani, Ebrahim 230, 232, 309
Manning, Chris 317, 326
Marcus, Gary 279, 326
Markert, Katja 301, 326
Masthoff, Judith 327
mathematics 71–5, 90, 93, 101–2, 127, 135, 141, 156, 200, 267, 297, 309, 316 mathematics: mathematical continuity 150–1, 303, 305, 308, 316 mathematical induction 75–6 mathematical

proof 102 vagueness in mathematics 71–7, 84–9
maxint 76, 90, 297
Mayr, Ernst 20, 26, 295, 326
McLeod, Ian 312, 326
measurement 1, 31–6, 38–9, 42–4, 46–7, 51–3, 71, 77, 82, 84–9, 102–3, 107, 127–8, 131, 143, 178–4, 191, 199–200, 210, 230, 236–7, 245, 250, 263–8, 275, 278–9, 282–92, 294–6, 302, 308, 315–7 see also error: margin of error; distance measurement 32–5, 70, 79, 87, 127, 130, 146, 281, 295 see also distance
median 42–3, 134
medicine 89, 154, 229, 294
Mellish, Chris 300, 330
membership: membership function 191–3, 232 membership graph 191–2, 231–2
memory: computer 82, 84, 222, 297 human 63, 175–6, 263, 312, 319
metamer 144
metaphor 115, 120, 189, 301
metre 32–5, 52, 79–80, 281
metric 37, 90, 247, 267–8, 275, 279, 294
metric system 279, 295
Mitkov, Ruslan 317, 326
model 12, 14, 61, 69, 122, 131, 135, 148–50, 160, 163–5, 169, 176, 179–83, 186, 189–90, 215, 221, 227–8, 234–5, 248–9, 253, 256, 258–9, 277, 282–4, 287, 301, 310–11, 315
model: complete model 163–4 partial model 161, 163–6, 170, 282
modus ponens, inference rule of 160
Moline, Jon 303, 326
Montague, Richard 93, 95–7, 101, 103, 106, 117, 124, 299, 326
morality 284
Morgenstern, Oskar 254, 331
Morreau, Michael 108, 296, 300, 311, 325–6
Morrison, Frances P. 323
mortality 37–8, 45, 52
Moxey, L. M. 109, 141, 326
Munro bagging 31
Munro, Hugh 31
Musil, Robert 61–2, 296
Muskens, Reinhard 174, 176, 307
Myerson, R. B. 311, 326